AF363297

Biorefinery

Biorefinery: Current Status, Challenges, and New Strategies

Editors

Kwang Ho Kim
Chang Geun Yoo

MDPI • Basel • Beijing • Wuhan • Barcelona • Belgrade • Manchester • Tokyo • Cluj • Tianjin

Editors
Kwang Ho Kim
Clean Energy Research Center
Korea Institute of Science and
Technology
Seoul
Korea, South

Chang Geun Yoo
Chemical Engineering
SUNY - ESF
Syracuse
United States

Editorial Office
MDPI
St. Alban-Anlage 66
4052 Basel, Switzerland

This is a reprint of articles from the Special Issue published online in the open access journal *Applied Sciences* (ISSN 2076-3417) (available at: www.mdpi.com/journal/applsci/special_issues/ biorefinery_challenges_strategies).

For citation purposes, cite each article independently as indicated on the article page online and as indicated below:

LastName, A.A.; LastName, B.B.; LastName, C.C. Article Title. *Journal Name* **Year**, *Volume Number*, Page Range.

ISBN 978-3-0365-1510-6 (Hbk)
ISBN 978-3-0365-1509-0 (PDF)

Contents

Editorial

Editorial on Special Issue "Biorefinery: Current Status, Challenges, and New Strategies"

Kwang Ho Kim [1],* and Chang Geun Yoo [2],*

1   Clean Energy Research Center, Korea Institute of Science and Technology, Seoul 02702, Korea
2   Department of Chemical Engineering, State University of New York College of Environmental Science and Forestry, Syracuse, NY 13210, USA
*   Correspondence: kwanghokim@kist.re.kr (K.H.K.); cyoo05@esf.edu (C.G.Y.)

Keywords: biorefinery; biomass fractionation; lignin valorization; bioproducts

Citation: Kim, K.H.; Yoo, C.G. Editorial on Special Issue "Biorefinery: Current Status, Challenges, and New Strategies". *Appl. Sci.* **2021**, *11*, 4674. https://doi.org/10.3390/app11104674

Received: 11 May 2021
Accepted: 17 May 2021
Published: 20 May 2021

Publisher's Note: MDPI stays neutral with regard to jurisdictional claims in published maps and institutional affiliations.

The overdependence on fossil fuels has raised concerns about global warming and the energy crisis, which has warranted significant research to find alternatives. Lignocellulosic biomass has attracted significant attention because it is renewable, abundant, and carbon-neutral [1]. The past decade has seen increasing research efforts to develop biomass-to-bioproducts processes. Despite the huge potential proven by recent biorefineries, the current bioeconomy still faces various technical challenges. In this respect, modern lignocellulosic biorefineries aim to develop more sustainable biomass conversion processes, tackling such challenges.

Fractionation of biomass components, typically the first stage in biomass conversion processes, remains an essential step to facilitate the separation of lignin and polysaccharides. In this Special Issue, Ahmed et al. [2] studied the effect of gamma-valerolactone (GVL)-assisted biomass fractionation on lignin extraction and cellulose digestion. GVL was recently viewed as a sustainable alternative in biomass defragmentation [3]. In this article, the authors reported that 80% aqueous GVL could effectively remove lignin and enhance enzymatic digestibility. Lee et al. [4] used a two-step pretreatment approach that includes an acid-catalyzed steam explosion followed by alkali-catalyzed organosolv treatment to separate cellulose-rich fraction and lignin from an empty fruit bunch.

In addition to the fractionation methods, this Special Issue covers various bio-products that can be produced from lignocellulosic biomass. For example, Kim et al. [5] explored a simultaneous conversion and extraction of furfural from hydrolysates using an aqueous biphasic system. They specifically focused on converting pentoses found in dilute acid hydrolysates into furfural, and the maximum yield of furfural was up to 94.6%. Jang et al. [6] presented the autohydrolysis of sweet sorghum bagasse to produce fermentable sugars and xylooligosaccharides (XOSs). Considering that the application areas of XOSs continue to expand, the production of XOSs from non-edible sources (i.e., lignocellulosic biomass) would be highly promising. Cho et al. [7] reported the catalytic conversion of α-pinene into high-density fuel candidates over stannic chloride molten salt hydrates ($SnCl_4 \cdot 5H_2O$), suggesting reaction mechanisms. Su et al. [8] summarized current metabolic engineering strategies that have been applied to biomass-to-isobutanol conversion processes. They also introduced recent advances in the production of isobutanol from various biomass feedstocks.

Biomass characterization is also an important area of modern biorefineries. Recently, diverse biomass processings, including chemical, thermal, biological, and hybrid processes, have been developed. It requires an in-depth characterization of biomass, understanding how those processes affect biomass structure and properties. Zhuang et al. [9] systematically used Fourier transform infrared (FTIR) analysis to characterize different types of biomass and their products. They successfully identified contaminants on the biomass surface. The information obtained from this study could help to prevent misunderstanding the FTIR analysis results of the processes biomass. Koo et al. [10] studied the drying effect of enzymatic hydrolysis of cellulose. They adopted the drying effect index (DEI) determined by analysis of porosity and crystallinity, to evaluate the effect of drying on the following processing. It was found that the drying effect was correlated with cellulose porosity, mainly due to the fiber hornification.

In recent biorefineries, there has been a shift from perceiving lignin as waste to viewing lignin as a potential material for value-added produces. Considerable efforts have been made to develop lignin conversion processes to produce specialty chemicals and polymeric materials to replace petroleum-based ones. More recently, attention has been placed in sectors, including the medical, electrochemical, and polymer sector, where lignin can be valorized. Yu and Kim wrote a review covering the recent research progress in lignin valorization, specifically focusing on medical, electrochemical, and 3D printing applications with technoeconomic insights [11].

Lastly, Harahap et al. [12] presented interesting research on technoeconomic evaluation of hand sanitizer production using oil palm empty fruit bunch (OPEFB)-based bioethanol. The COVID-19 pandemic increased the demand for ethanol as the primary ingredient of hand sanitizers. In this article, they evaluated the technoeconomic feasibility of hand sanitizer production using bioethanol produced from OPEFB. The results clearly suggest that the production of hand sanitizer from bioethanol is economically viable and can be implemented at a tolerable price as an alternative application for second-generation bioethanol.

Funding: This research received no external funding.

Institutional Review Board Statement: Not applicable.

Informed Consent Statement: Not applicable.

Data Availability Statement: Not applicable.

Conflicts of Interest: The authors declare no conflict of interest.

References

1. Yu, O.; Yoo, C.G.; Kim, C.S.; Kim, K.H. Understanding the Effects of Ethylene Glycol-Assisted Biomass Fractionation Parameters on Lignin Characteristics Using a Full Factorial Design and Computational Modeling. *ACS Omega* **2019**, *4*, 16103–16110. [CrossRef] [PubMed]
2. Ahmed, M.A.; Lee, J.H.; Raja, A.A.; Choi, J.W. Effects of Gamma-Valerolactone Assisted Fractionation of Ball-Milled Pine Wood on Lignin Extraction and Its Characterization as Well as Its Corresponding Cellulose Digestion. *Appl. Sci.* **2020**, *10*, 1599. [CrossRef]
3. Shuai, L.; Questell-Santiago, Y.M.; Luterbacher, J.S. A mild biomass pretreatment using γ-valerolactone for concentrated sugar production. *Green Chem.* **2016**, *18*, 937–943. [CrossRef]
4. Lee, J.H.; Ahmed, M.A.; Choi, I.G.; Choi, J.W. Fractionation of Cellulose-Rich Products from an Empty Fruit Bunch (EFB) by Means of Steam Explosion Followed by Organosolv Treatment. *Appl. Sci.* **2020**, *10*, 835. [CrossRef]
5. Kim, J.H.; Cho, S.M.; Choi, J.H.; Jeong, H.; Lee, S.M.; Koo, B.; Choi, I.G. A Simultaneous Conversion and Extraction of Furfural from Pentose in Dilute Acid Hydrolysate of Quercus mongolica Using an Aqueous Biphasic System. *Appl. Sci.* **2021**, *11*, 163 [CrossRef]
6. Jang, S.K.; Jung, C.D.; Yu, J.H.; Kim, H. Environmentally Friendly Approach for the Production of Glucose and High-Purity Xylooligosaccharides from Edible Biomass Byproducts. *Appl. Sci.* **2020**, *10*, 8119. [CrossRef]
7. Cho, S.M.; Choi, J.H.; Kim, J.H.; Koo, B.; Choi, I.G. Catalytic Conversion of alpha-Pinene to High-Density Fuel Candidates Over Stannic Chloride Molten Salt Hydrates. *Appl. Sci.* **2020**, *10*, 7517. [CrossRef]
8. Su, Y.D.; Zhang, W.W.; Zhang, A.L.; Shao, W.J. Biorefinery: The Production of Isobutanol from Biomass Feedstocks. *Appl. Sci.* **2020**, *10*, 8222. [CrossRef]
9. Zhuang, J.S.; Li, M.; Pu, Y.Q.; Ragauskas, A.J.; Yoo, C.G. Observation of Potential Contaminants in Processed Biomass Using Fourier Transform Infrared Spectroscopy. *Appl. Sci.* **2020**, *10*, 4345. [CrossRef]
10. Koo, B.; Jo, J.; Cho, S.M. Drying Effect on Enzymatic Hydrolysis of Cellulose Associated with Porosity and Crystallinity. *Appl. Sci.* **2020**, *10*, 5545. [CrossRef]

11. Yu, O.; Kim, K.H. Lignin to Materials: A Focused Review on Recent Novel Lignin Applications. *Appl. Sci.* **2020**, *10*, 4626. [CrossRef]
12. Harahap, A.F.P.; Panjaitan, J.R.H.; Curie, C.A.; Ramadhan, M.Y.A.; Srinophakun, P.; Gozan, M. Techno-Economic Evaluation of Hand Sanitiser Production Using Oil Palm Empty Fruit Bunch-Based Bioethanol by Simultaneous Saccharification and Fermentation (SSF) Process. *Appl. Sci.* **2020**, *10*, 5987. [CrossRef]

Article

Effects of Gamma-Valerolactone Assisted Fractionation of Ball-Milled Pine Wood on Lignin Extraction and Its Characterization as Well as Its Corresponding Cellulose Digestion

Muhammad Ajaz Ahmed [1], Jae Hoon Lee [2], Arsalan A. Raja [3] and Joon Weon Choi [1,*]

[1]   Institute of Green-Bio Science and Technology, Seoul National University, Pyeongchang 232-916, Korea; m.ajaz@snu.ac.kr
[2]   Department of Forest Sciences, Seoul National University, Seoul 151-921, Korea; tirchonail@snu.ac.kr
[3]   Department of Chemical Engineering, University of Hafr Al-Batin, Hafr Al-Batin 31991, Saudi Arabia; arsalanr@uohb.edu.sa
*   Correspondence: cjw@snu.ac.kr

Received: 31 January 2020; Accepted: 25 February 2020; Published: 28 February 2020

Abstract: Gamma-valerolactone (GVL) was found to be an effective, sustainable alternative in the lignocellulose defragmentation for carbohydrate isolation and, more specifically, for lignin dissolution. In this study, it was adapted as a green pretreatment reagent for milled pinewood biomass. The pretreatment evaluation was performed for temperature (140–180 °C) and reaction time (2–4 h) using 80% aqueous GVL to obtain the highest enzymatic digestibility of 92% and highest lignin yield of 33%. Moreover, the results revealed a positive correlation ($R^2 = 0.82$) between the lignin removal rate and the crystallinity index of the treated biomass. Moreover, under the aforementioned conditions, lignin with varying molecular weights (150–300) was obtained by derivatization followed by reductive cleavage (DFRC). 2D heteronuclear single quantum coherence nuclear magnetic resonance (2D-HSQC-NMR) spectrum analysis and gel permeation chromatography (GPC) also revealed versatile lignin properties with relatively high β-O-4 linkages (23.8%–31.1%) as well as average molecular weights of 2847–4164 with a corresponding polydispersity of 2.54–2.96, indicating this lignin to be a heterogeneous feedstock for value-added applications of biomass. All this suggested that this gamma-valerolactone based pretreatment method, which is distinctively advantageous in terms of its effectiveness and sustainability, can indeed be a competitive option for lignocellulosic biorefineries.

Keywords: pinewood; green pretreatment; enzymatic hydrolysis; lignin structural features

1. Introduction

In the context of the non-sustainable supply of fossil fuels, lignocelluloses have emerged as one potentially suitable option to theoretically swamp the existing petroleum-based fuels [1]. These lignocellulosic substrates are highly likely to shape the future of the bioeconomy given their sustainable supply, relative abundance and readily collectible nature [2]. These biomasses, in their typical native conformation, are composed of sugar polymers of glucose and xylose and are surrounded by a protective sheath of lignin [3]. This renewable feedstock is a rich source for the production of fuels, commodity chemicals and even biomaterials [4]. However, in their native form, certain features associated with their structural integrity make them unresponsive to any external attacks [5]. For example, the complexity of the cell wall structure, degree of polymerization, cellulose crystallinity, extent of lignification and compositional heterogeneity even within one species of biomass, etc. [6,7] are such major factors. Therefore, pretreatment, to make biomass available for subsequent processing, has been envisaged as an unavoidable step in biomass refining processes [8]. This fractionation step, in principle, disrupts the biomass native structure, exposes the sugar polymers for the subsequent enzymatic process and makes lignin available for easy extraction [9]. There have been a fair deal of studies involving a variety of pretreatment strategies including acids, alkaline reagents, hydrothermal pretreatment, organosolv pretreatment, ionic liquids, steam explosion and even biological pretreatments [10]. These pretreatments have their pros and cons [11]. For example, acidic pretreatments, along with steam explosion and hydrothermal ones, dissolve mostly the xylose portion and keep lignin within a biomass web [12]. On the other hand, ionic liquids also have significant effects on biomass fractionation, but their burdensome cost prohibits their commercial-scale applications [13]. Moreover, biological treatments are associated with prolonged processing times, making them unfavorable for biorefining processes [14]. Organosolv treatments, as compared with other pretreatment schemes, cause considerable alterations in the structural features of biomass [15]. These treatments disrupt the biomass by creating internal slits for enhanced bioconversion efficiency. Moreover, from these solvent pulping strategies, lignin can be recovered with significant characteristic features [16]. This isolated lignin, with adequate purity having low polysaccharides, is highly desired for making lignin-based compounds with tailored properties. Recently, gamma-valerolactone (GVL) has emerged as one such sustainable green solvent for biomass fractionation [17]. This organic solvent is a self-sustained non-toxic product formed within a lignocellulosic biorefinery and is capable enough to extract lignin at near-neutral solvent conditions [18]. Typically, GVL/acid/water has been mostly applied for an enhanced delignification rate with improved cellulose digestion [19]. As it is miscible with water, it induces better hydrolytic effects for biomass to dissolve out the lignin portion [20]. This reagent has also been applied with synergetic effects of metal salts, microwaves and acids [21,22]. Therefore, in this study, we adapted a GVL assisted fractionation of pinewood biomass to study its effects on the lignin isolation and its characteristic features along with the enzymatic digestion of corresponding carbohydrate fractions. Formally, 80% aqueous GVL was employed along with ball-milled pinewood into stainless steel reactors and was treated for time intervals of 2–4 h with a temperature range of 140–180 °C.

2. Materials and Methods

2.1. Preparation of Biomass

Pinewood biomass was chosen as a model substrate in this study. It was collected from Gangwon Province, South Korea. Prior to being employed for experimentation, it was air-dried, chopped and finally ground to a particle size of 60 mesh. It was subsequently methanol extracted for 24 h and was then air-dried at a temperature of 60 °C. The compositional analysis of this methanol extracted biomass was determined according to two-step acid hydrolysis of the National Renewable Energy Laboratory's (NREL) standard analytical procedure. Its composition was revealed to be: glucan 44%, xylan 7%, mannan 13%, Klason lignin 28% and 4% moisture contents. All the chemicals used in this study were of analytical grade and were employed in experimentation without any further modification.

2.2. GVL Assisted Fractionation of the Biomass

The pinewood biomass was subjected to milling for 20 h (on the basis of initial laboratory trials data not shown) using a laboratory ball mill prior to being used in the pretreatment experiments. Later on, lab-scale specially designed stainless-steel reactors, with a working volume of 300 mL, were used in the pretreatment experiments. The milled pinewood was loaded into the reactors at a solids loading of 5% (20:1 liquid to solid ratio). A measured amount of gamma-valerolactone was added, and the slurry was mixed well to near complete homogenization prior to heating on an oil bath. For all the experiments, a pH value of 6–7 was already present, and no external addition of any acidic reagent was made during the designated pretreatment heating time. After fractionation, the reactors were quenched on an ice bath to cool down to room temperature. Afterward, the biomass slurry was subjected to filtration with a Buchner funnel and vacuum pump assembly to collect the solid residue and the dissolved lignin fractions as filtrate. The liquid fractions were subjected to lignin separation via the water precipitation method. For this, a specified volume of DI water (4 times the experimental volume) was added to all the recovered liquid fractions and was left overnight. It was then separated by centrifugation, freeze-dried and saved in the refrigerator for any further experiments. The lignin yield was calculated on the basis of the initial lignin contents of the raw biomass on a weight basis. The corresponding filtered solid residue was washed and dried in an oven overnight for subsequent processing.

2.3. Enzymatic Saccharification

For enzymatic hydrolysis, GVL treated oven-dried biomass was loaded into Erlenmeyer flasks at 5% solid loading with 50 mM sodium citrate and citric acid buffer solution according to one of our previous studies [23]. The enzymatic blend of Cellic C-Tec 2 (Novozymes, Denmark), derived from *Trichoderma reesei*, with a dose of 30 FPU/g of biomass was used. This enzymatic saccharification was performed at 50 °C and 200 rpm for 72 h in a shaking incubator, and the samples were taken periodically.

2.4. Compositional Analysis and Biosugar Quantification

To determine the biomass composition, a two-step acid hydrolysis protocol adapted from the National Renewable Energy Laboratory (NREL) was employed [24]. Briefly, dried treated biomass was first subjected to 72% (v/v) H_2SO_4 hydrolysis at room temperature, then the second hydrolysis was done with autoclaving at 121 °C for 1 h. After cooling at room temperature, the liquid was filtered out and prepared for HPLC analysis, whereas the solid residue was dried overnight for lignin quantification. The monomeric sugar fractions were quantified by HPLC equipped with an RID detector (Thermoscientific Ultimate 3000) and a Carbosep CHO-87 P column (Concise Solutions, USA). The column temperature was set at 65 °C, and DI water was used as a mobile phase with a flow rate of 0.6 mL/min.

2.5. Statistical Analysis

We used regression analysis to interpret the correlation of enzymatic digestibility and XRD with lignin yield. The stated R^2 values have been reported on the basis of linear regression to clarify the explanatory roles of enzymatic digestibility and XRD on the lignin yield, respectively. This analysis was performed with the help of SigmaPlot version 10.0.

2.6. Characterization of Isolated Lignin and Fractionated Biomass

The recovered lignin fractions were analyzed by gel permeation chromatography (GPC), derivatization followed by reductive cleavage (DFRC) and 2D heteronuclear single quantum coherence nuclear magnetic resonance (2D-HSQC-NMR) spectrum according to our previous study [25]. To determine the weight average (Mw) and number average (Mn) molecular weights, each isolated fraction was first subjected to acetylation in pyridine/acetic anhydride (1:1) solution and was periodically shaken after being maintained at a temperature of 105 °C for 2 h. After this, almost 20 mg of this acetylated lignin was dissolved in 2 mL of tetrahydrofuran and then analyzed using a GPC, Agilent 1200, USA (equipped with an ultraviolet detector (UV) on a PL-gel 5 mm Mixed-c. column, calibrated with polystyrene standards), and THF was used as a mobile phase while it was flowed at a rate of 1 mL/min. For DFRC analysis, a previous classical approach was adapted in which the extracted lignin samples, 20 mg for each sample, were dissolved in 4 mL AcOH and 1 mL AcBr solution and kept at 50 °C for 1 h with rigorous periodic shaking. Afterward, the residue was obtained by evaporating the mixture on a rotary evaporator. To this residue, 2 mL of dioxane/acetic acid/water (5:4:1, v/v/v) was added, and subsequently, zinc dust (50 mg) was also added with stirring and then the whole solution was kept for 1 h at room temperature. Later on, after the intended time, a precise amount of the internal standard tetracosane was added into the solution, which was extracted by the mixture with CH_2Cl_2 and saturated NH_4Cl. After this, the pH of the aqueous phase was dropped down to 3 by adding 3% HCl, and then finally, the organic layer was separated. The water phase was extracted twice with CH_2Cl_2, and the combined CH_2Cl_2 fractions were dried over $MgSO_4$, and ultimately, the filtrate was evaporated under reduced pressure. The residue was then subjected to acetylation employing 0.5 mL each of acetic anhydride and pyridine for at least 1 h with periodic shaking. All the residual matter was then filtered using syringe filters and finally subjected to quantification by GC-MS (Agilent 7890B USA). For 2D-HSQC-NMR, the lignin samples were prepared using 20 mg of lignin dissolved in 0.75 mL DMSO-D6, and then the analysis data were acquired with a Bruker AVANCE 600 spectrometer (Bruker, Germany). Raw and pretreated pinewood samples were characterized in terms of their crystallinity indices as well. For this, the crystallinity index (CrI) was determined by measuring X-ray diffraction (XRD) using a powder X-ray diffractometer (Bruker, D8 Advance with DAVINCI Germany). A sample was scanned at a rate of 2°/min in the 2θ range of 3°–60°, and the CrI was estimated according to Equation (1).

$$\mathrm{CrI}\ (\%) = [(I_{\mathrm{Crystalline}} - I_{\mathrm{Amorphous}})/I_{\mathrm{Crystalline}}] \times 100\% \tag{1}$$

where $I_{\mathrm{crystalline}}$ = intensity at 22.26° and $I_{\mathrm{amorphous}}$ = intensity at 15.72°.

3. Results and Discussion

3.1. Effects of Pretreatment Severity on the Chemical Composition of Biomass

Temperature, in general, is a critical parameter in chemical reactions, and this is also the case for lignocellulosic biomass pretreatment. To observe the effects of GVL fractionation under different thermal conditions on the milled pinewood composition, a reaction was conducted for 2–4 h in the presence of 80% aqueous GVL with temperature varying from 120 °C to 160 °C.

The chemical composition of the treated biomass along with solid recovery are listed in Table 1. The solid residue, left after the GVL treatment, ranged from 86% to 50%, mainly exhibiting a counter-wise behavior to the treatment severity. This high biomass dissolution has already been found in a similar GVL assisted fractionation of biomass [26].

Table 1. Effects of pretreatment conditions on the chemical composition of pinewood biomass.

Sample	GVL Extraction Conditions		Solid Recovery (%)	Solid Composition (%)				Solution pH after Complete Reaction
	Time (hours)	Temperature (°C)		Glucan	Xylan	Mannan	Lignin	
Pinewood	2	140	86	42	5	14	29	6.0
	4	140	80	38	6	12	30	5.9
	2	160	70	37	4	15	26	5.5
	4	160	65	34	10	12	25	5.2
	2	180	60	28	9	10	24	4.0
	4	180	50	25	8	12	30	3.0

The biosugars, glucan and mannan, were mainly not altered due to this severity showing only a scant variation in their composition. Xylan, however, was dissolved a bit more than the glucan and mannan. This xylan composition, as a typical amorphous regions dissolution trend, was also supported by the low pH value that was a direct function of the treatment severity. This pH variation ranged from 6 at 140:2 h to 3 at the harshest conditions of 180 °C for a treatment operation of 4 h. The glucan and mannan were also affected at a high pH and high thermal environment, indicating that GVL was also effective in dissolving the cellulose portion, although only partly. This biosugar recovery, despite the harsh processing conditions, was quite high. For example, it was substantially higher than a previous Fenton-based study in which almost half of the glucan was lost [27].

3.2. Lignin Extraction from Ball-Milled Pinewood

In order to find out the effects of treatment conditions on the lignin yield, the lignin extraction efficiency, based on initial raw lignin contents, was also determined and is shown in Table 2. Briefly, the filtrate obtained from all of the six performed experiments under our described conditions was first added to 800 mL of DI water and left overnight to settle down the lignin as precipitates. Afterward, these fractions were centrifuged using a laboratory centrifuge machine. The supernatant was decanted off carefully, and the leftover solid fractions were subjected to freeze-drying for 24 h. These freeze-dried lignin samples were used for their characterization and further analysis. The lignin yield, in a somewhat expected manner, showed a gradual trend in comparison with the treatment severity. For example, starting from less harsh treatment conditions of 140:2, it was quite a low yield, with only 3.0% lignin. However, as the thermal severity was increased over a long retention time, the yield was correspondingly increased to 33%. This trend showed that the ball-milled biomass was dissolved in GVL and it was likely to separate out the acetyl contents to create an acidic environment as the treatment severity was prolonged. The highest lignin yield of 33% was obtained at 180:4, the extreme conditions in our case. Interestingly, the biomass solid recovery and the corresponding pH value of the final solution was around 50% and 3, respectively. This shows that there was indeed some acidic environment due to harsher treatment conditions.

Table 2. Lignin yield and corresponding glucose release from treated biomass.

Sample	GVL Extraction Conditions		Lignin Yield (%)
	Time (hours)	Temperature (°C)	
	2	140	3.0
	4	140	6.5
Pinewood	2	160	18
	4	160	27
	2	180	29
	4	180	33

3.3. Enzymatic Hydrolysis of Solid Fractions

In order to observe the GVL assisted treatment efficacy, the cellulose digestion of the treated solid residues as well as the untreated ones was performed, and the results are delineated in Figure 1. As expected, the untreated biomass showed a very low enzymatic saccharification yield of nearly 25%, mainly due to its intact and smooth surface providing less effective enzymatic absorption sites. However, for the treated biomass fractions, the glucose release and the corresponding enzymatic digestibility rose in a gradual way, exhibiting that treatment severity had a positive impact on glucose release. Initially, the yield was not that high (almost 60% for 140–160 °C); however, it increased in a dramatic fashion (92%) when the thermal conditions went beyond 160 °C viz (180 °C:4 h). The lignin was also removed when the treatment conditions were prolonged, so it can be anticipated that, in our case, the biomass web deconstruction, along with the removal of lignin, provided enhanced accessibility for enzymatic penetration. From here, it can be anticipated that the cellulose digestion was drastically improved for the ball-milled pinewood biomass with GVL pretreatment. This treatment operation, due to its biomass dissolution properties, disrupted the biomass native matrix and exposed inner surfaces, which escalated the subsequent enzymatic saccharification. From here, it can be conclusively suggested that this treatment can be one suitable option for lignocellulosic ethanol production.

3.4. Correlation of Enzymatic Hydrolysis with Delignification and Crystallinity Index

Since there was a significant lignin removal during the GVL treatment process, we also determined the crystallinity index of the treated and the untreated biomass samples to observe any correlation with enzymatic digestibility. As shown in Figure 2, both the crystallinity indices along with the delignification showed a positive correlation with the enzymatic digestibility. The high value of R^2 also proves this good agreement. From here, it can be easily concluded that the governing factor for heightened enzymatic hydrolysis in our case is both the lignin removal as well as the crystallinity index. A similar trend was also observed in a previous study where they found an $R^2 = 0.72$ for lignin removal and enzymatic hydrolysis [28]. Interestingly, a previous study found a negative correlation of crystallinity with the enzymatic digestibility [19], but in our study, a contrary result was found, which shows that the governing factor for enhanced glucose yield, at least in our case, is both the crystallinity index and the lignin removal.

Enzymatic Digestibility VS Time

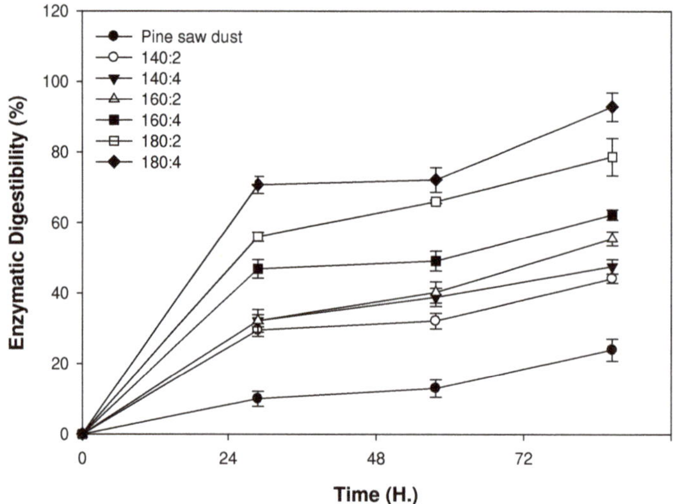

Figure 1. Digestibility and time graph.

Correlation of lignin yiled with enzymatic digestibility & XRD

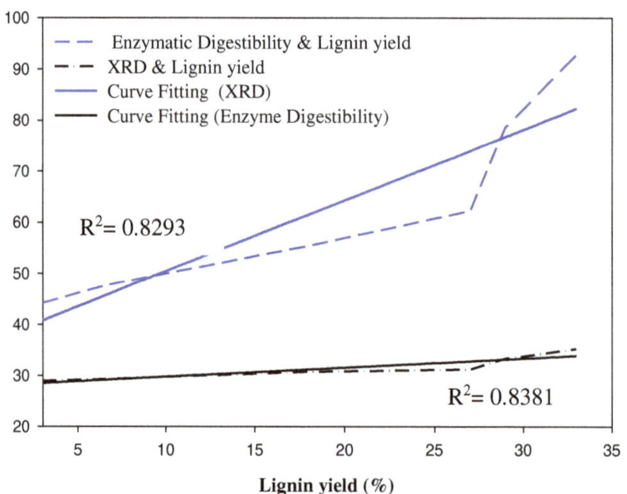

Figure 2. Correlation of enzymatic digestibility and XRD with lignin yield.

3.5. Gel Permeation Chromatography

For the extracted lignin fractions, molecular weight analysis and chromatograms employing the GPC method were performed to more precisely understand their fragmentation during GVL fractionation. The results for this analysis are shown in Table 3, along with the weight average (M_w) and number average (M_n) molecular weights as well as the corresponding polydispersity (M_w/M_n).

Table 3. Results of molecular weights and polydispersity.

Sample	Mn	Mw	(Mw/Mn) P
Pine MWL	5157	10660	2.06
GVL Lignin 140:2	4164	12331	2.96
GVL Lignin 140:4	4942	11913	2.41
GVL Lignin 160:2	4575	13162	2.87
GVL Lignin 160:4	3897	10165	2.60
GVL Lignin 180:2	3279	7939	2.42
GVL Lignin 180:4	2847	7245	2.54

As it is quite obvious in the table, the molecular weights ranged from 12,331 to 7245 gm/mol, which is relatively high compared to a classical lignin, i.e., milled wood lignin (MWL). In one previous study where GVL was employed as a pretreatment reagent, lignin also exhibited heterogeneous and diverse properties due to the harsher process conditions [26]. From here, it is presumable that the lignin underwent condensation during this GVL treatment. In regards to polydispersity, as can be seen from the table, the value is in the range of 2.54–2.96 and is representative of a narrow margin, which shows a relative heterogeneity of lignin macromolecules. However, this high value of polydispersity shows that there are large amounts of low molecular weight species present from the isolated lignin fractions. This same phenomenon is also supported by the behavior of the chromatograms, which is illustrated in Figure 3.

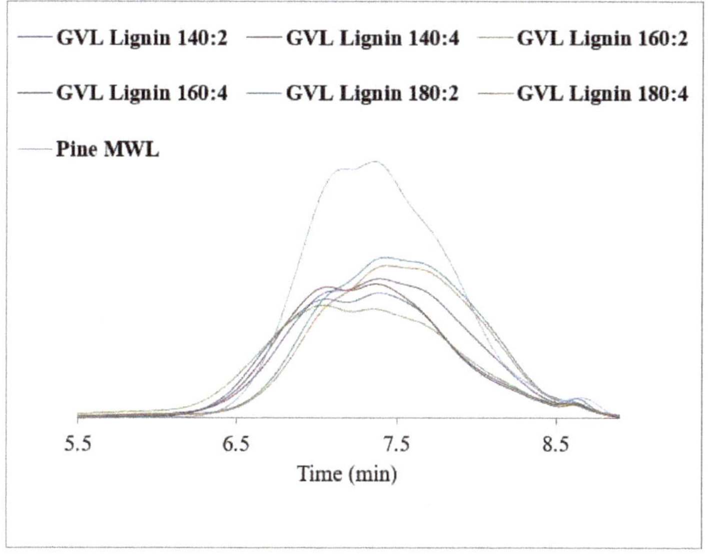

Figure 3. Gel permeation chromatography (GPC) chromatograms.

3.6. DFRC

For the sake of structural investigation of the obtained lignins, the DFRC method was applied via its typical selective β-aryl ether structural degradation. The fractions were studied by GC-MS, and the results are shown in Table 4. As can be seen in the table values, each lignin fraction represents only the G type of units due to the typical natural G-unit composition of pinewood biomass. The values for DFRC results range from 300 to 150 μmol/gm of lignin for GVL treatment conditions, whereas the value of 810 represents an MWL value for very high G-units. It is obvious from the table that as the treatment severity increases, there is a clear, meaningful, decreasing trend in the DFCR values, showing that the thermal conditions have led to the degradation of the lignin molecules. From here, it can be suggested that the isolated lignin fractions, however, were scant in comparison to MWL but were lower in their β-aryl ether units.

Table 4. Derivatization followed by reductive cleavage (DFRC)—GC/MS results.

Sample	G Units	S Units	Total
Pine MWL	810.6 ± 34.9	-	810.6 ± 34.9
GVL Lignin 140:2	300 ± 35.8	-	300 ± 35.8
GVL Lignin 140:4	250 ± 36.8	-	250 ± 36.8
GVL Lignin 160:2	242 ± 46.0	-	242 ± 46.0
GVL Lignin 160:4	210 ± 10.4	-	242 ± 10.4
GVL Lignin 180:2	196 ± 24.7	-	196 ± 24.7
GVL Lignin 180:4	150 ± 48.1	-	150 ± 48.1

3.7. 2D-HSQC-NMR Spectra

For 2D-HSQC-NMR, the lignin samples were prepared using 20 mg of lignin dissolved in 0.75 mL DMSO-D6, and then the analysis data were acquired with a Bruker AVANCE 600 spectrometer (Bruker, Germany). Meanwhile, the relative quantification and the corresponding assignments of A_α, B_α and C_α are tabulated in Table 5, and their corresponding NMR spectra are delineated and shown in Figure 4. It is quite obvious that the GVL extracted lignin, in comparison with MWL, was slightly lower in β–O–4 contents, ranging from 23.8% at more severe conditions to 31.1% at rather mild conditions. However, the other linkages of β- β and β-5 showed somewhat comparable values with MWL. In one previous study, 58.2% of β–O–4 linkages were reported when they extracted lignin from cotton stalks with 80% aqueous GVL but at 170 °C and employing H_2SO_4 as a catalyst [29]. Our lower values might be due to the crude form of the extracted lignin fractions.

Table 5. Quantification (%) of various structures in four selected lignin samples.

	β–O–4	β- β	β-5
Pine MWL	43.5	2.0	16.1
GVL Lignin 140:2	28.8	3.5	11.6
GVL Lignin 140:4	31.1	3.6	13.2
GVL Lignin 160:2	26.9	3.0	13.0
GVL Lignin 160:4	23.8	3.8	12.0

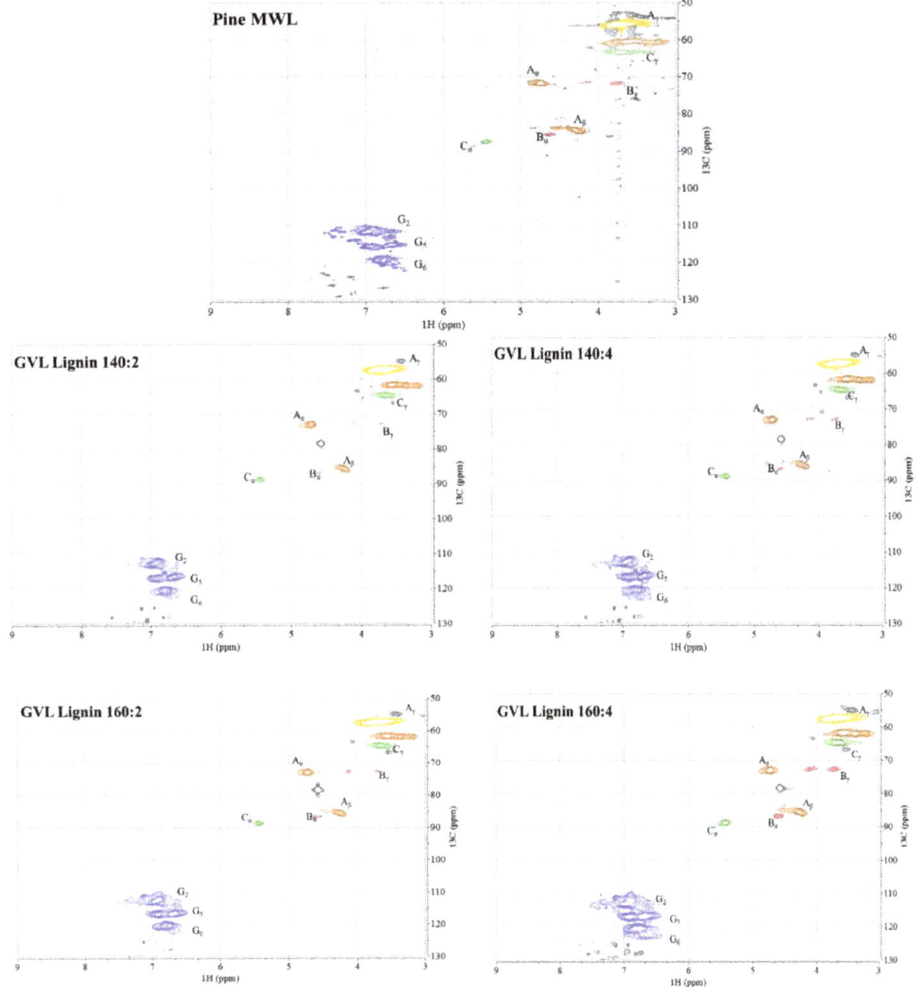

Figure 4. 2D-HSQC-NMR results for milled wood lignin (MWL) and four selected samples (140:2, 140:4, 160:2 and 160:4).

4. Conclusions

Gamma-valerolactone was found to be an efficient green reagent to deconstruct the native structure of milled pinewood biomass. This reagent, with only 80% aqueous concentration, together with the synergetic effect of heat, caused considerable compositional changes along with the high dissolution of lignin contents—up to 33%—and drastically enhanced the enzymatic digestibility of the corresponding polysaccharide fractions by 92%. This fractionation approach, as a green and sustainable alternative, seems to be a potential addition to already adapted organosolv pretreatment strategies for lignocellulosic biomass. Thus, it can be suggested as one potential non-toxic option for the lignocellulosic biorefining processes.

Author Contributions: Conceptualization, A.A.R.; Data curation, M.A.A.; Formal analysis, J.H.L.; Investigation, M.A.A.; Supervision, J.W.C. All authors have read and agreed to the published version of the manuscript.

Funding: This research was funded by the Basic Science Research Program (NRF-2019R1A2C2086328) and the Technology Development Program to Solve Climate Changes (2017M1A2A2087627) of the National Research Foundation funded by the Ministry of Science and ICT.

Conflicts of Interest: The authors declare no conflict of interest.

References

1. Ubando, A.T.; Felix, C.B.; Chen, W. Biorefineries in circular bioeconomy: A comprehensive review. *Bioresour. Technol.* **2019**, 122585. [CrossRef] [PubMed]
2. Venkata Mohan, S.; Dahiya, S.; Amulya, K.; Katakojwala, R.; Vanitha, T.K. Can circular bioeconomy be fueled by waste biorefineries—A closer look. *Bioresour. Technol. Rep.* **2019**, *7*, 100277. [CrossRef]
3. Araújo, D.; Vilarinho, M.; Machado, A. Effect of combined dilute-alkaline and green pretreatments on corncob fractionation: Pretreated biomass characterization and regenerated cellulose film production. *Ind. Crop. Prod.* **2019**, *141*. [CrossRef]
4. Jin, J.; Yu, B.; Shi, Z.; Wang, C.; Chong, C. Lignin-based electrospun carbon nanofibrous webs as free-standing and binder-free electrodes for sodium ion batteries. *J. Power Sources* **2014**, *272*, 800–807. [CrossRef]
5. Khoshnevisan, B.; Shafiei, M.; Rajaeifar, M.A.; Tabatabaei, M. Biogas and bioethanol production from pinewood pre-treated with steam explosion and N-methylmorpholine-N-oxide (NMMO): A comparative life cycle assessment approach. *Energy* **2016**, *114*, 935–950. [CrossRef]
6. Kumar, B.; Bhardwaj, N.; Agrawal, K.; Chaturvedi, V.; Verma, P. Current perspective on pretreatment technologies using lignocellulosic biomass: An emerging biorefinery concept. *Fuel Process. Technol.* **2020**, *199*. [CrossRef]
7. Tian, S.Q.; Zhao, R.Y.; Chen, Z.C. Review of the pretreatment and bioconversion of lignocellulosic biomass from wheat straw materials. *Renew. Sustain. Energy Rev.* **2018**, *91*, 483–489. [CrossRef]
8. Jafari, Y.; Amiri, H.; Karimi, K. Acetone pretreatment for improvement of acetone, butanol, and ethanol production from sweet sorghum bagasse. *Appl. Energy* **2016**, *168*, 216–225. [CrossRef]
9. Stücker, A.; Schütt, F.; Saake, B.; Lehnen, R. Lignins from enzymatic hydrolysis and alkaline extraction of steam refined poplar wood: Utilization in lignin-phenol-formaldehyde resins. *Ind. Crop. Prod.* **2016**, *85*, 300–308. [CrossRef]
10. Kumari, D.; Singh, R. Pretreatment of lignocellulosic wastes for biofuel production: A critical review. *Renew. Sustain. Energy Rev.* **2018**, *90*, 877–891. [CrossRef]
11. Hassan, S.S.; Williams, G.A.; Jaiswal, A.K. Moving towards the second generation of lignocellulosic biorefineries in the EU: Drivers, challenges, and opportunities. *Renew. Sustain. Energy Rev.* **2019**, *101*, 590–599. [CrossRef]
12. Solarte-Toro, J.C.; Romero-García, J.M.; Martínez-Patiño, J.C.; Ruiz-Ramos, E.; Castro-Galiano, E.; Cardona-Alzate, C.A. Acid pretreatment of lignocellulosic biomass for energy vectors production: A review focused on operational conditions and techno-economic assessment for bioethanol production. *Renew. Sustain. Energy Rev.* **2019**, *107*, 587–601. [CrossRef]
13. Sorn, V.; Chang, K.L.; Phitsuwan, P.; Ratanakhanokchai, K.; Dong, C.D. Effect of microwave-assisted ionic liquid/acidic ionic liquid pretreatment on the morphology, structure, and enhanced delignification of rice straw. *Bioresour. Technol.* **2019**, *293*, 121929. [CrossRef] [PubMed]
14. Zabed, H.M.; Akter, S.; Yun, J.; Zhang, G.; Awad, F.N.; Qi, X.; Sahu, J.N. Recent advances in biological pretreatment of microalgae and lignocellulosic biomass for biofuel production. *Renew. Sustain. Energy Rev.* **2019**, *105*, 105–128. [CrossRef]
15. Cebreiros, F.; Clavijo, L.; Boix, E.; Ferrari, M.D.; Lareo, C. Integrated valorization of eucalyptus sawdust within a biorefinery approach by autohydrolysis and organosolv pretreatments. *Renew Energy* **2020**, *149*, 115–127. [CrossRef]
16. Ramakoti, B.; Dhanagopal, H.; Deepa, K.; Rajesh, M.; Ramaswamy, S.; Tamilarasan, K. Solvent fractionation of organosolv lignin to improve lignin homogeneity: Structural characterization. *Bioresour. Technol. Rep.* **2019**, *7*, 100293. [CrossRef]
17. Tan, X.; Zhang, Q.; Wang, W.; Zhuang, X.; Deng, Y.; Yuan, Z. Comparison study of organosolv pretreatment on hybrid pennisetum for enzymatic saccharification and lignin isolation. *Fuel* **2019**, *249*, 334–340. [CrossRef]
18. Ye, L.; Han, Y.; Feng, J.; Lu, X. A review about GVL production from lignocellulose: Focusing on the full components utilization. *Ind. Crop. Prod.* **2020**, *144*, 112031. [CrossRef]
19. Sun, S.N.; Chen, X.; Tao, Y.H.; Cao, X.F.; Li, M.F.; Wen, J.L.; Nie, S.X.; Sun, R.C. Pretreatment of Eucalyptus urophylla in γ-valerolactone/dilute acid system for removal of non-cellulosic components and acceleration of enzymatic hydrolysis. *Ind. Crop. Prod.* **2019**, *132*, 21–28. [CrossRef]

20. Jia, L.; Qin, Y.; Wen, P.; Zhang, T.; Zhang, J. Alkaline post-incubation improves cellulose hydrolysis after Γ-valerolactone/water pretreatment. *Bioresour. Technol.* **2019**, *278*, 440–443. [CrossRef]

21. Jin, L.; Yu, X.; Peng, C.; Guo, Y.; Zhang, L.; Xu, Q.; Zhao, Z.K.; Liu, Y.; Xie, H. Fast dissolution pretreatment of the corn stover in gamma-valerolactone promoted by ionic liquids: Selective delignification and enhanced enzymatic saccharification. *Bioresour. Technol.* **2018**, *270*, 537–544. [CrossRef] [PubMed]

22. Li, X.; Liu, Q.; Si, C.; Lu, L.; Luo, C.; Gu, X.; Liu, W.; Lu, X. Green and efficient production of furfural from corn cob over H-ZSM-5 using γ-valerolactone as solvent. *Ind. Crop. Prod.* **2018**, *120*, 343–350. [CrossRef]

23. Ahmed, M.A.; Rehman, M.S.U.; Terán-Hilares, R.; Khalid, S.; Han, J.I. Optimization of twin gear-based pretreatment of rice straw for bioethanol production. *Energy Convers. Manag.* **2017**, *141*, 120–125. [CrossRef]

24. Sluiter, J.B.; Ruiz, R.O.; Scarlata, C.J.; Sluiter, A.D.; Templeton, D.W. Compositional analysis of lignocellulosic feedstocks. 1. Review and description of methods. *J. Agric. Food Chem.* **2010**, *58*, 9043–9053. [CrossRef] [PubMed]

25. Kim, J.Y.; Park, S.Y.; Lee, J.H.; Choi, I.G.; Choi, J.W. Sequential solvent fractionation of lignin for selective production of monoaromatics by Ru catalyzed ethanolysis. *RSC Adv.* **2017**, *7*, 53117–53125. [CrossRef]

26. Angelini, S.; Ingles, D.; Gelosia, M.; Cerruti, P.; Pompili, E.; Scarinzi, G.; Cavalaglio, G.; Cotana, F.; Malinconico, M. One-pot lignin extraction and modification in γ-valerolactone from steam explosion pre-treated lignocellulosic biomass. *J. Clean. Prod.* **2017**, *151*, 152–162. [CrossRef]

27. Jung, Y.H.; Kim, H.K.; Park, H.M.; Park, Y.C.; Park, K.; Seo, J.H.; Kim, K.H. Mimicking the Fenton reaction-induced wood decay by fungi for pretreatment of lignocellulose. *Bioresour. Technol.* **2015**, *179*, 467–472. [CrossRef]

28. Ko, J.K.; Bak, J.S.; Jung, M.W.; Lee, H.J.; Choi, I.G.; Kim, T.H.; Kim, K.H. Ethanol production from rice straw using optimized aqueous-ammonia soaking pretreatment and simultaneous saccharification and fermentation processes. *Bioresour. Technol.* **2009**, *100*, 4374–4380. [CrossRef]

29. Wu, M.; Liu, J.K.; Yan, Z.Y.; Wang, B.; Zhang, X.M.; Xu, F.; Sun, R.-C. Efficient recovery and structural characterization of lignin from cotton stalk based on a biorefinery process using a γ-valerolactone/water system. *RSC Adv.* **2016**, *6*, 6196–6204. [CrossRef]

Article

Fractionation of Cellulose-Rich Products from an Empty Fruit Bunch (EFB) by Means of Steam Explosion Followed by Organosolv Treatment

Jae Hoon Lee [1], Muhammad Ajaz Ahmed [2], In-Gyu Choi [1] and Joon Weon Choi [2,*]

[1] Department of Forest Sciences, Seoul National University, Seoul 151-921, Korea; tirchonail@snu.ac.kr (J.H.L.); cingyu@snu.ac.kr (I.-G.C.)
[2] Institute of Green-Bio Science and Technology, Seoul National University, Pyeongchang 232-916, Korea; m.ajaz@snu.ac.kr
* Correspondence: cjw@snu.ac.kr; Tel.: +82-33-339-5840

Received: 12 December 2019; Accepted: 16 January 2020; Published: 24 January 2020

Abstract: In this study an empty fruit bunch (EFB) was subjected to a two-step pretreatment to defragment cellulose-rich fractions as well as lignin polymers from its cell walls. First pretreatment: acid-catalyzed steam explosion (ACSE) pretreatment of EFB was conducted under the temperature range of 180–220 °C and residence time of 5–20 min. The ACSE-treated EFB was further placed into the reactor containing 50% aq. ethanol and NaOH as a catalyst and heated at a temperature of 160 °C for 120 min for the second pretreatment: alkali-catalyzed organosolv treatment (ACO). The mass balance and properties of treated EFB were affected by the residence time. The lowest yield of a solid fraction was obtained when the residence time was kept at 15 min. Xylose drastically decreased, especially under the ACSE pretreatment. However, the crystallinity of cellulose increased by increasing the severity factor of the pretreatment and was 47.8% and 57% udner the most severe conditions. The organosolv lignin fractions also showed the presence of 14 major peaks via their pyrolysis-GC analysis. From here, it can be suggested that this kind of pretreatment can indeed be one potential option for lignocellulosic pretreatment.

Keywords: Biomass; two-step pretreatment; steam explosion; organosolv treatment; empty fruit bunch

---

1. Introduction

Over the recent decades, the utilization of empty fruit bunch (EFB) to produce renewable energy source has been studied extensively. This, EFB typically, is a leftover of the fresh fruit bunches after their fruit harvesting. Moreover, it is the major downstream residue from the Southeast Asian palm oil industry. This industry, recently, has produced 70 million metric tons of crude palm oil (CPO) just within one production year 2017–2018 amounting their production even up to 35% of the total production of edible vegetable oils [1]. Theoretically, palm oil plantations can produce more than 100 million metric tons of EFB since 1.42–1.88 tons of EFB are generated for every one ton of CPO produced [2]. However, unfortunately, most of these EFB is dumped and pose serious environmental problems. Therefore, it is highly desirable to utilize these biomasses for sustainable energy production.

Biochemical conversion of lignocelluloses involving chemical pretreatment followed by enzymatic hydrolysis and finally anaerobic fermentation is a well-established production line for lignocellulosic bioethanol production. In this classical approach, due to an inherited complex cell wall structure, effective fractionation of major lignocellulosic biomass components (cellulose, hemicellulose, and lignin) is crucial for effective subsequent processing. Undoubtedly, there have been several pretreatment studies involving EFB to deconstruct its native web into its constituents. In this regard, steam explosion, which uses high-pressure steam to breakdown biomass structure, has been widely adapted strategy [3]. The effects of several factors like biomass particle size, moisture, pressure, residence time, reaction temperature and catalysts on the steam explosion were consistently investigated for several decades [4]. Previously, Choi et al. used alkaline reagent (3% NaOH at 160 °C) to catalyze steam explosion pretreatment and dissolved out C5 sugar-based components as well as lignin components from biomass [5]. Their yields, after enzymatic hydrolysis, for glucose and xylose heightened up to 93% and 78% respectively. In another study, the steam explosion of rose gum tree (Eucalyptus grandis) was carried out under the temperature range of 200–210 °C and residence time of 2–5 min after impregnation with 0.087% and 0.175% (*w/w*) H_2SO_4 [6]. Their results showed that 0.175% (*w/w*) H_2SO_4-impregnated chips at 210 °C for 2 min was the optimal condition for hemicellulose extraction. Today steam explosion is seen as one of the most widely employed and cost-effective process for lignocellulose pretreatment, yet the process still needs to be improved to realize industrialization [7]. In addition to this steam explosion strategy, another effective fractionation scheme for lignocelluloses is organosolv treatment, so-called organosolvation. This process, in principle, extracts lignin from lignocellulose with organic solvents, such as alcohols, acetone and organic peracids [8]. Previous studies have shown considerable effects of organosolv treatment (diethylene glycol) of rice straw and EFB as compared to soda pulping by Gonzalez et al. [9] keeping in view the heterogeneous yet complex cell wall structure of EFB, it is highly likely to investigate an appropriate pretreatment strategy for commercial production of biofuels [10]. Therefore, it has to be investigated the proper pretreatment process of EFB. In this study we adapted a two-step pretreatment strategy to fractionate EFB biomass into its sugar and non-sugar components. For this two-step pretreatments, EFB was first subjected to acid-catalyzed steam explosion (ACSE) followed by alkali-catalyzed organosolv treatment (ACO). Pretreatment efficacy was evaluated in terms of yields, crystallinity, and chemical composition of each solid fraction.

2. Materials and Methods

2.1. Feedstock Analysis

As raw biomass for this study, EFB (≤6% moisture) was provided by the Korea Institute of Science and Technology, in a setting of 0.5–2 mm fiber. Holocellulose, lignin, and ash content in the EFB sample were analyzed according to the National Renewable Energy Laboratory (NREL) standard procedures [11–13]. Carbon, hydrogen, and nitrogen contents were measured by using a CHNS-932 analyzer (LECO Corp., St. Joseph, MI, USA) and inorganic elements were analyzed by inductively coupled plasma emission spectroscopy (iCAP7400 Duo, Thermo Fisher Scientific, Waltham, MA, USA) [14]. Thermogravimetric analysis was performed under an inert atmosphere (50 mL/min N_2 flow) at a heating rate of 10 °C/min up to 800 °C with a TGA/DSC 3+ (METTLER TOLEDO, Columbus, OH, USA) [14]. All the results with the corresponding values are listed in Table 1.

Table 1. Chemical and thermal characteristics of empty fruit bunch.

Properties	Value
Elemental analysis (wt%)	
Carbon	45.4 ± 0.4
Hydrogen	5.0 ± 0.2
Nitrogen	0.6 ± 0.0
Oxygen [1]	49.0 ± 0.2

Table 1. *Cont.*

Properties	Value
Component analysis (wt%)	
Holocellulose	70.6 ± 0.2
Arabinose	6.1
Galactose	3.1
Glucose	39.0
Xylose	18.8
Lignin	18.0 ± 0.2
Extractives	4.6 ± 0.1
Ash	6.2 ± 0.3
Inorganic element (ppm)	
Aluminum	1760
Calcium	2170
Iron	2680
Magnesium	790
Potassium	11,930
Phosphorus	750
Silicon	760
Thermogravimetric analysis	
Volatiles (wt%)	80.5
Char (wt%)	19.5
Temperature at max. degradation rate (°C)	318.8

[1] By difference.

2.2. ACSE-First Pretreatment

Initially, we screened out experimental conditions based on our initial trials and proceeded accordingly as described below for the pretreatment. The complete fractionation processes are delineated in Figure 1. Briefly, for ACSE treatment, EFB was first washed to remove inorganic elements using a method suggested by Moon et al. [15] Later on it was impregnated in distilled water for 120 min with slow stirring, and finally the feedstock was dried in an oven at 75 °C. It was then impregnated in dilute sulfuric acid (0.5, 1, 2 wt%) for 4 h to make it ready for steam explosion pretreatment. The effect of dilute acid concentration on impregnation was examined by the component analysis (holocellulose, lignin, and sugar) [12].

For ACSE experiments, autoclave reactors with 2 L high pressure steam were loaded first with 150 g of raw EFB biomass samples on dry basis as a control. The final temperature in the reactor was applied to reach in the interval of 180–220 °C and the residence time was 5–15 min. After the intended reaction time, a valve was opened to release the pressure immediately. Solid fractions (E-S) from the steam explosion was washed with distilled water, and the resulting aqueous fraction (E-L) was also collected. Then, acid impregnated EFB, with the same solids loading, was steam-treated under the temperature of 220 °C and 20 min of residence time, the most severe conditions that could be stably maintained with the reactor. After the reaction and cooling, solid fraction (EA-S), and aqueous fraction (EA-L) were also collected.

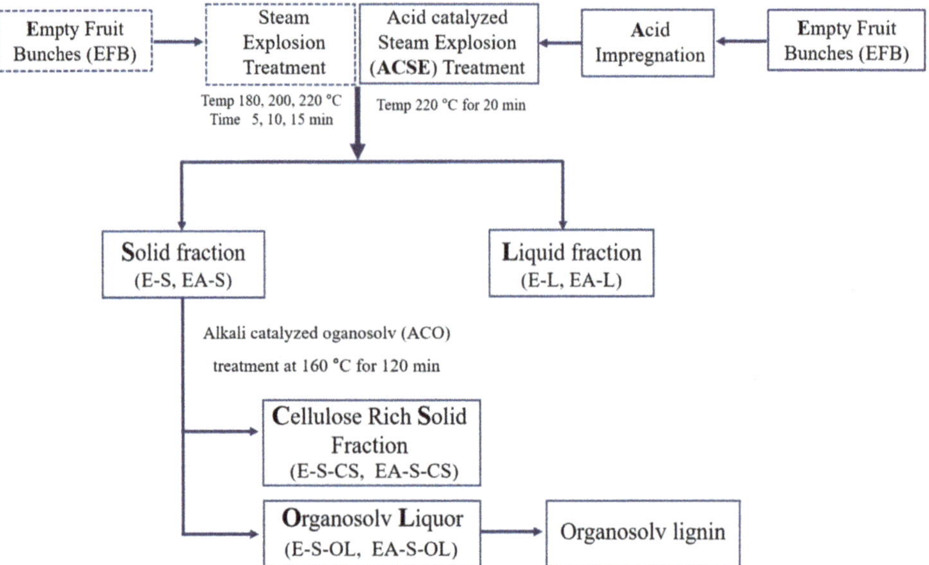

Figure 1. Schematic diagram of the two-step fractionation process of empty fruit bunches (EFB) by (acid-catalyzed) steam treatment followed by alkali-catalyzed organosolv treatment.

2.3. ACO Second Pretreatment

Organosolv treatment experiments, as a 2nd treatment for both the E-S & EA-S solid fractions obtained from 1st treatment, were performed as follows: both the solid fractions were placed, one by one, in a solvent-circulating reactor with a solvent of a 50:50% (*w/w*) ethanol:water mixture maintaining a (10:1 (*w/w*) solvent:solid ratio) along with a 2 wt% of NaOH as a catalyst. The reaction was carried out at a temperature of 160 °C and 120 min residence time. After the desired reaction time, cellulose-rich solid fractions were collected and labelled as E-S-CS and EA-S-CS and liquid fractions were labelled as E-S-OL and EA-S-OL corresponding to E-S and EA-S biomass samples, respectively. The organosolv lignin was precipitated from E-S-OL and EA-S-OL by acidification with a 2 M of hydrochloric acid maintaining pH = 2. The precipitated organosolv lignin was freeze-dried and collected as a powder and saved for further experiments.

2.4. Analysis

Yields of each solid fraction were calculated using the following equation:

$$\text{Yield (\%)} = \text{weight of a solid fraction (g)/weight of dry feedstock (g)} \times 100$$

Holocellulose, lignin, and sugar content in each solid fraction were analyzed by the same process mentioned in Section 2.1. X-ray powder diffraction patterns (XRD) were analyzed by a Bruker D8 Advance with DAVINCI using Cu Kα radiation (λ = 1.5418 Å), operated at 40 kV and 40 mA with a scan speed of 0.5 s/min in a range of 2–50 degrees (2 thetas). Chemical compounds in liquid fraction were quantified and qualified using gas chromatography-mass spectrometry systems (5975C Series GC/MSD System, Agilent Technologies, Santa Clara, CA, USA) [14]. Organosolv lignin was analyzed by a coil-type CDS Pyroprobe 5000 (CDS Analytical Inc., Oxford, PA, USA) [16].

3. Results and Discussion

3.1. Mass Balance

3.1.1. Effect of Residence Time and Temperature on Fractionation

The effects of residence time and temperature on the fractionation process were studied at various intervals (residence time 5, 10, 15 min and temperature 180, 200, 220 °C). Figure 2 shows the mass distribution of solid fractions obtained from EFB at various temperatures and residence time conditions.

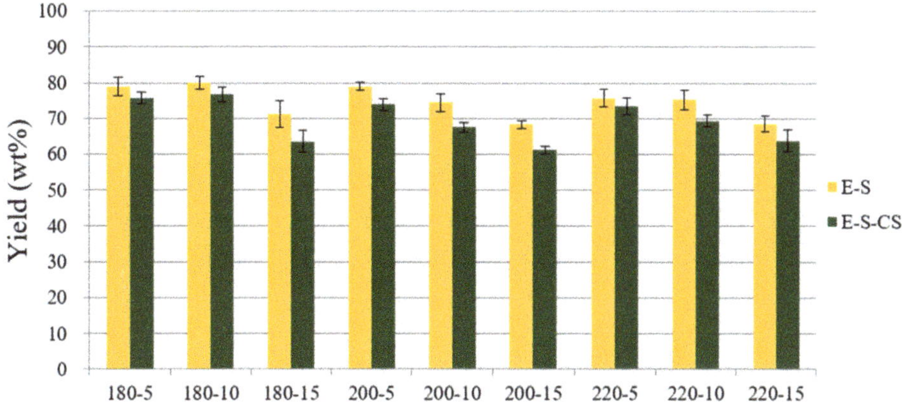

Figure 2. Yields of the solid fractions from steam explosion and organosolv treatment of EFB under different temperatures and residence times.

Mass composition of the fraction was affected by residence time. At all temperature conditions, E-S from the steam explosion, and its subsequent, E-S-CS from organosolv treatment decreased with increasing residence time. For example, at 200 °C and in the range of 5–15 min, yields of E-S and E-S-CS decreased from 79.0 wt% and 73.8 wt% to 68.3 wt% and 61.2 wt%, respectively. The lowest yield was obtained when the residence time was kept for 15 min, regardless of reaction temperatures. The effect of temperature was not much clear as that of residence time. Under the same residence time condition, the yield of E-S and E-S-CS decreased only 3–8 wt%. The combined effect of temperature (T) and residence time (t) can be expressed as severity factor $(\log_{10}(R_0))$ $[R_0 = t \times \exp((T-100)/14.75))$ [17]. The severity factor obtained ranged from 3.1 to 4.7. There was a scant negative correlation between the severity factor and the solid fraction yield, showing that increasing temperature and residence time results in the removal of compounds from solid fractions.

In general, both mechanical and chemical effects take place when an acetyl group derived from hemicellulose acts as an acid at high temperature during the steam explosion [18]. These effects play important roles in the hydrothermal degradation of holocellulose. Hemicellulose and lignin are isolated by the high temperature in the steam condition. Organosolv treatment, however, on the other hand, solubilize lignin and provide holocellulose by using organic solvent mixture and inorganic catalysts [7]. Therefore, holocellulose and lignin compositions of solid fractions were also analyzed to examine the fractionation efficiency of the process.

Holocellulose and lignin contents of E-S and E-S-CS were calculated for each experiment (Table 2). Percentage of holocellulose composition decreased with increasing severity of steam explosion. A maximum of 24.1 wt% of holocellulose (220-15) was dissolved in E-L. On the contrary, the percentage of lignin composition increased with increasing severity factor. It can be anticipated that lignin melts under a high temperature and pressure, then condensed with other compounds like extractives or ash while cooling. Solid fractions which have undergone organosolvation showed an unremarkable change of holocellulose contents. In addition, a considerable amount of lignin content still remained in the solids. Maximum 7.5% of lignin (220-15) was solubilized in 50% ethanol mixture. Lignin fractionation efficiency of organosolv treatment from EFB was falling short of our expectation because of its low lignin removal. These results showed EFB presents a high level of recalcitrance to the fractionation process. Thus, dilute acid pretreatment to level down the EFB recalcitrance had considerable effects and the results are discussed in Section 3.1.2.

Table 2. Component analysis of pretreated EFB solid fractions under different temperatures and residence times.

		Holocellulose (wt%)	Lignin (wt%)
Raw EFB		70.6 ± 0.2	18.0 ± 0.2
	180-5	70.2 ± 0.1	23.3 ± 0.0
	180-10	69.3 ± 0.2	22.9 ± 0.1
	180-15	70.8 ± 0.2	25.1 ± 0.2
	200-5	71.2 ± 0.1	24.2 ± 1.0
E-S	200-10	69.9 ± 0.1	27.9 ± 0.8
	200-15	71.6 ± 0.4	29.3 ± 0.8
	220-5	68.3 ± 0.1	29.4 ± 0.8
	220-10	67.5 ± 0.0	33.7 ± 0.1
	220-15	67.9 ± 0.3	34.6 ± 0.2
	180-5	80.8 ± 0.3	20.1 ± 0.1
	180-10	80.3 ± 0.2	20.2 ± 0.1
	180-15	81.1 ± 0.2	21.6 ± 0.3
	200-5	83.6 ± 0.1	20.6 ± 0.1
E-S-CS	200-10	80.2 ± 0.2	22.8 ± 0.5
	200-15	81.5 ± 0.1	22.9 ± 0.9
	220-5	83.8 ± 0.2	21.5 ± 0.6
	220-10	74.9 ± 0.8	24.6 ± 0.5
	220-15	76.2 ± 0.1	22.8 ± 0.2

E-S: solid fraction obtained from steam explosion treatment of EF. E-S-CS: cellulose rich solid fraction obtained from alkali catalyzed organosolv treatment of E-S fraction.

3.1.2. Effect of Dilute Acid Impregnation on Fractionation

Acid impregnation has been considered to play a vital role for steam explosion treatment [4,19]. Keeping this in view, the effects of dilute acid on the fractionation process was studied with the feedstock impregnation in dilute sulfuric acid (0.5, 1, 2 wt%) for 4 h. Steam explosion process was further operated at residence time of 20 min and temperature of 220 °C, which showed the highest severity score so that the effect of dilute acid concentration can be compared clearly. Figure 3 shows the mass distribution of solid fractions obtained from EFB at various acid concentrations.

Mass composition of EA-S did not affect by the acid concentration clearly until the concentration reaches 2.0%. Moreover, at 0.5 and 1.0% dilute acid concentrations, EA-S from ACSE also showed no remarkable changes in the mass distribution. In the range of 0%–1.0% dilute acid concentration, yields of EA-S decreased from 73.0 to 70.8 wt%. However, EA-S-CS from ACO decreased with an increase in the acid concentration. In the same range of 0%–1.0% dilute acid concentration, EA-S-CS decreased from 59.0 to 45.9 wt%. At 2.0% dilute acid concentration, EA-S and EA-S-CS decreased with an increase in the acidity.

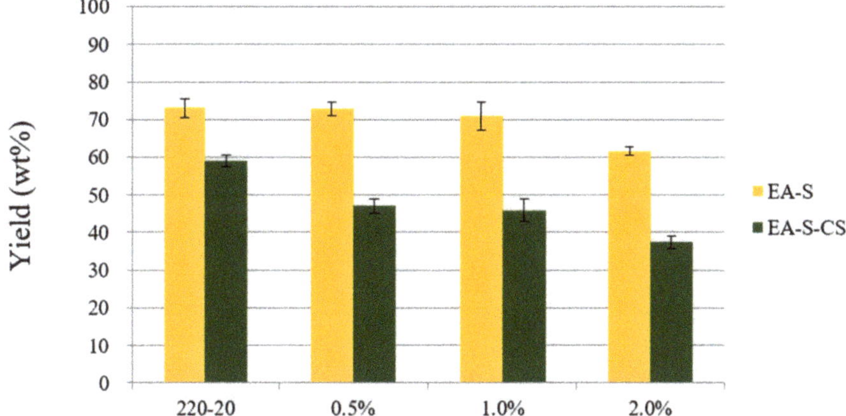

Figure 3. Yields of the solid fractions from acid catalyzed steam explosion (ACSE) and alkali-catalyzed organosolv treatment (ACO) of EFB.

Table 3 shows the holocellulose and lignin contents of EA-S and EA-S-CS calculated for each experimental run. The total amount of holocellulose harshly decreased with increasing dilute acid concentration. At 0.5% dilute acid concentration, 41.3 and 26.8 wt% of holocellulose still remained in EA-S and EA-S-CS, respectively. When the concentration increased to 2.0%, however, only 28.1 and 16.0 wt% of holocellulose remained in EA-S and EA-S-CS, respectively. Lignin exhibited somewhat different behavior in mass decrease. The amount of lignin did change slightly by the steam treatment. Hwoever, after ACO, 4.3%–6.7% of lignin was removed from the solid fractions. The removal efficiency of lignin between 0.5% and 2.0% acid concentration respectively. Similar studies were reported for the dilute acid treatment of rice straw [20] and corn stover [21]. However, the proper acid concentration or removal efficiencies of hemicellulose and lignin were different by feedstock properties. Since the purpose of this experiments is to lower only hemicellulose and lignin contents in EFB, so a high acid concentration was not adapted to avoid the loss of holocellulose contents.

Table 3. Component analysis of acid pretreated EFB solid fractions.

		Holocellulose (wt%)	Lignin (wt%)
Raw EFB		70.6 ± 0.2	18.0 ± 0.2
EA-S	220-20	69.0 ± 0.2	24.4 ± 0.8
	0.5%	56.7 ± 0.2	33.7 ± 0.2
	1.0%	52.0 ± 0.1	38.0 ± 0.1
	2.0%	45.6 ± 0.1	45.9 ± 0.8
EA-S-CS	220-20	64.1 ± 0.1	23.2 ± 0.8
	0.5%	56.5 ± 0.4	25.3 ± 0.4
	1.0%	49.3 ± 0.3	27.9 ± 0.4
	2.0%	42.8 ± 0.1	30.1 ± 0.3

EA-S: solid fraction obtained from acid catalyzed steam explosion treatment of acid impregnated EFB. EA-S-CS: cellulose rich solid fraction obtained from alkali catalyzed organosolv treatment of EA-S fraction.

3.1.3. Sugars Quantification

Tables 4 and 5 show the total sugar contents of each product fractions (EA-S, EA-S-CS, EA-L, and EA-S-OL). Glucose content in the solid fractions showed no significant change while lignin content increased compared to the raw material. It means that the glucose in the EFB took limited damage by the steam explosion process and the organosolv treatment. It is possible that long reaction times increased the accumulation of degradation by-products and mass losses by volatilization [22]. As the acid concentration increased, however, the amount of glucose in the solid steadily decreased. When the acid concentration increased to 2.0%, the glucose amount decreased from 415 to 380 mg/g fraction. In contrast, concentration of glucose in EA-L dropped with 0.5 wt% acid pretreatment but increased by increasing acid concentration. Xylosyl groups were the most affected hemicellulose components during acid impregnation and the pretreatment process. The amount of xylose in the solid drastically decreased with increasing dilute acid concentration. As a result, concentration of xylose in the liquid fraction increased to twice of the control (10.1–21.9 mg/mL liquid). At 2.0% dilute acid concentration, more than 70% of the xylan in the EFB was hydrolyzed during two-step pretreatments. Arabinosyl, galactosyl, and mannosyl residues were removed from the solids by steam explosion with more than 0.5% acid concentration (Table 5).

Table 4. Sugar content of pretreated EFB solid fractions under different temperatures and residence times.

		Total Sugar (mg/g Fraction)					Lignin (mg/g Fraction)
		Glucose	Xylose	Arabinose	Galactose	Mannose	
EA-S	220-20	483	170	23	10	4	244
	0.5%	433	104	-	-	-	337
	1.0%	421	76	-	-	-	380
	2.0%	373	33	-	-	-	459
EA-S-CS	220-20	447	161	20	8	5	232
	0.5%	426	89	-	-	-	253
	1.0%	405	50	-	-	-	279
	2.0%	380	48	-	-	-	301

Table 5. Sugar content of acid pretreated EFB liquid fractions.

		Total Sugar (mg/mL Liquid)					Lignin (mg/mL Liquid)
		Glucose	Xylose	Arabinose	Galactose	Mannose	
EA-L	220-20	12.2	10.1	3.6	0.3	-	n.d.
	0.5%	0.3	13.4	0.7	0.3	-	n.d.
	1.0%	1.4	20.6	0.8	0.4	0.2	n.d.
	2.0%	7.9	21.9	1.1	0.7	1.0	n.d.
EA-S-OL	220-20	-	2.2	3.0	-	0.4	38.2
	0.5%	0.8	3.2	-	-	-	19.0
	1.0%	0.1	1.4	-	-	-	20.5
	2.0%	-	2.4	-	-	-	12.2

EA-L: liquid fraction obtained from acid catalyzed steam explosion treatment of acid impregnated EFB. EA-S-OL: organosolv liquor obtained from alkali catalyzed organosolv treatment of EA-S fraction.

3.2. Cellulose Crystallinity

The change of cellulose crystallinity as a result of the steam explosion and organosolv treatment was determined by analyzing the x-ray diffraction patterns (Figure 4 and Table 6). By the XRD peak height method, crystallinity index was calculated from the ratio of the height of the 002 peaks (22.7°) and the height of the minimum (18.3°) between the 002 and the 101 peaks [23]. The crystallinity of EFB fiber decreased due to the steam explosion and organosolv treatment. The lowest crystallinity was shown in E-S-CS under the condition of 180-10 (35.9%). The crystallinity of EFB increased by increasing the severity factor. This might be caused by the removal of the amorphous structure of cellulose or components like hemicellulose.

Cellulose crystallinity after the dilute acid impregnation followed by two-step pretreatment was determined by analyzing the X-ray diffraction patterns. The crystallinity of EFB fiber also decreased by both ACSE process and ACO. There was little change of crystallinity, however, between the raw and organosolv treatment.

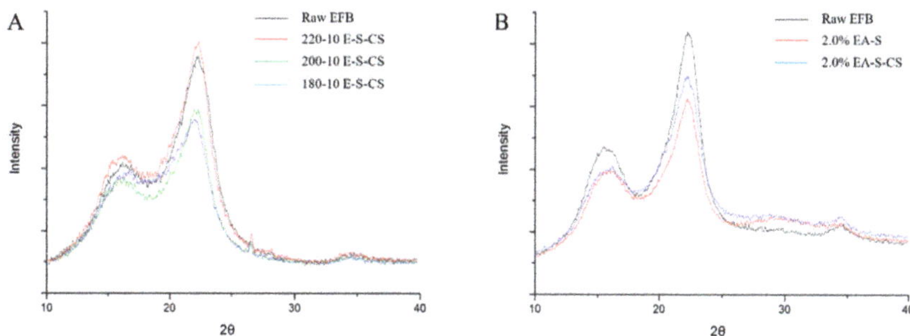

Figure 4. X-ray diffraction of (**A**) E-S-CS and (**B**) EA-S and follow-up EA-S-CS.

Table 6. X-ray diffraction of (**A**) E-S-CS and (**B**) EA-S and follow-up EA-S-CS.

			Crystallinity Index (%)
	Raw		58.3
		180-10	35.9
E-S-CS		200-10	47.8
		220-10	60.0
	EA-S 2.0%		47.8
	EA-S-CS 2.0%		57.0

3.3. Chemical Properties of Organosolv Lignin

Qualitative and quantitative analysis of obtained organosolv lignins from E-S-OL were performed via pyrolysis-GC-MS analysis, and the results are listed in Table 7. Organosolv lignin obtained under steam explosion conditions of 200 °C and 5–10 min exhibited more than 14 peaks. Phenolic compounds, such as toluene (1), phenol (2), guaiacol (5), syringol (8), and isoeugenol (10), were the major compounds. As shown in Table 7, the number of chemical compounds and the overall peak size in the organosolv lignin decreased by increasing residence time. This might be explained by the fact that the lignin compounds can be condensed in harsher conditions and less decomposed by heat.

Table 7. Qualitative and quantitative analysis of chemical compounds in the organosolv lignin from E-S-OL.

No.	Compound	Concentration (Area/IS Area)	
		220-5	**220-10**
1	Toluene	0.45	0.15
2	Phenol	4.66	1.67
3	*o*-Cresol	0.44	0.16
4	*p*-Cresol	0.54	0.19
5	Guaiacol	0.51	0.33
6	Creosol	0.53	0.25
7	2-Methoxy-4-vinylphenol	0.40	0.12
8	Syringol-	1.01	0.61
9	3,5-Dimethoxy-4-hydroxytoluene	1.05	0.33
10	Isoeugenol	0.30	0.10
11	2,6-Dimethyl-3-(methoxymethyl)-benzoquinone	0.55	0.12
12	(E)-4-Propenylsyringol	0.68	0.19
13	n-Hexadecanoic acid	1.07	0.36
14	Octadecanoic acid	0.33	0.15
		12.50	4.72

4. Conclusions

Two-step pretreatment, ACSE followed by ACO of EFB, was performed under a temperature range of 180–220 °C and residence time of 5–20 min after impregnation with 0.5%–2.0% dilute sulfuric acid. The mass balance and the chemical composition of pretreated EFB fractions were noticeably affected by the reaction temperature, residence time, and acid concentration during the steam explosion process. Moreover, glucose and xylose contents decreased under the acid-catalyzed pretreatment conditions. All the analyses together with the XRD peaks showed that the crystallinity of cellulose increased by increasing severity factor. In this experiment, the effect of two-step pretreatment was in no way remarkable but showed potential to advance it. Further study is needed to sort out the optimal conditions for this two-step EFB pretreatment to conclusively suggest its potential for EFB fractionation.

Author Contributions: Conceptualization, J.W.C.; data curation, J.H.L.; investigation, J.H.L.; methodology, J.H.L. and J.W.C.; writing—original draft preparation, J.H.L.; writing—review and editing, M.A.A., I.-G.C. and J.W.C.; supervision, J.W.C. All authors have read and agreed to the published version of the manuscript.

Funding: This research was funded by the Basic Science Research Program (NRF-2019R1A2C2086328) and the Technology Development Program to Solve Climate Changes (2017M1A2A2087627) of the National Research Foundation funded by the Ministry of Science and ICT.

Conflicts of Interest: The authors declare no conflict of interest.

References

1. United States Department of Agriculture Economics, Statistics and Market Information System. Oilseeds: World Markets and Trade. Available online: https://downloads.usda.library.cornell.edu/usda-esmis/files/tx31qh68h/sq87bv06k/9z903025j/oilseed-trade-09-12-2018.pdf (accessed on 21 December 2018).

2. Pleanjai, S.; Gheewala, S.H.; Garivait, S. Environmental evaluation of biodiesel production from palm oil in a life cycle perspective. *Asian J. Energy Environ.* **2007**, *8*, 15–32.

3. Shrotri, A.; Kobayashi, H.; Fukuoka, A. Catalytic conversion of structural carbohydrates and lignin to chemicals. *Adv. Catal.* **2017**, *60*, 59–123.

4. Brownell, H.; Yu, E.; Saddler, J. Steam-explosion pretreatment of wood: Effect of chip size, acid, moisture content and pressure drop. *Biotechnol. Bioeng.* **1986**, *28*, 792–801. [CrossRef] [PubMed]

5. Choi, W.-I.; Park, J.-Y.; Lee, J.-P.; Oh, Y.-K.; Park, Y.C.; Kim, J.S.; Park, J.M.; Kim, C.H.; Lee, J.-S. Optimization of NaOH-catalyzed steam pretreatment of empty fruit bunch. *Biotechnol. Biofuels* **2013**, *6*, 170. [CrossRef] [PubMed]

6. Emmel, A.; Mathias, A.L.; Wypych, F.; Ramos, L.P. Fractionation of eucalyptus grandis chips by dilute acid-catalysed steam explosion. *Bioresour. Technol.* **2003**, *86*, 105–115. [CrossRef]

7. Alvira, P.; Tomas-Pejo, E.; Ballesteros, M.; Negro, M.J. Pretreatment technologies for an efficient bioethanol production process based on enzymatic hydrolysis: A review. *Bioresour. Technol.* **2010**, *101*, 4851–4861. [CrossRef] [PubMed]

8. Zhao, X.; Cheng, K.; Liu, D. Organosolv pretreatment of lignocellulosic biomass for enzymatic hydrolysis. *Appl. Microbiol. Biotechnol.* **2009**, *82*, 815. [CrossRef] [PubMed]

9. Gonzalez, M.; Canton, L.; Rodriguez, A.; Labidi, J. Effect of organosolv and soda pulping processes on the metals content of non-woody pulps. *Bioresour. Technol.* **2008**, *99*, 6621–6625. [CrossRef] [PubMed]

10. Lee, J.H.; Moon, J.G.; Choi, I.-G.; Choi, J.W. Study on the thermochemical degradation features of empty fruit bunch on the function of pyrolysis temperature. *J. Korean Wood Sci. Technol.* **2016**, *44*, 350–359. [CrossRef]

11. Sluiter, A.; Hames, B.; Ruiz, R.; Scarlata, C.; Sluiter, J.; Templeton, D. *Determination of Ash in Biomass*; Laboratory Analytical Procedure (LAP) Technical Report NREL/TP-510-42622; National Renewable Energy Laboratory: Golden, CO, USA, 2005.

12. Sluiter, A.; Hames, B.; Ruiz, R.; Scarlata, C.; Sluiter, J.; Templeton, D.; Crocker, D. *Determination of Structural Carbohydrates and Lignin in Biomass*; Laboratory Analytical Procedure (LAP) Technical Report NREL/TP-510-42618; National Renewable Energy Laboratory: Golden, CO, USA, 2008.

13. Sluiter, A.; Ruiz, R.; Scarlata, C.; Sluiter, J.; Templeton, D. *Determination of Extractives in Biomass*; Laboratory Analytical Procedure (LAP) Technical Report NREL/TP-510-42619; National Renewable Energy Laboratory: Golden, CO, USA, 2005.

14. Lee, J.H.; Hwang, H.; Moon, J.; Choi, J.W. Characterization of hydrothermal liquefaction products from coconut shell in the presence of selected transition metal chlorides. *J. Anal. Appl. Pyrol.* **2016**, *122*, 415–421. [CrossRef]

15. Moon, J.; Lee, J.H.; Hwang, H.; Choi, I.-G.; Choi, J.W. Effect of inorganic constituents existing in empty fruit bunch (EFB) on features of pyrolysis products. *J. Korean Wood Sci. Technol.* **2016**, *44*, 629–638. [CrossRef]

16. Kim, J.-Y.; Lee, J.H.; Park, J.; Kim, J.K.; An, D.; Song, I.K.; Choi, J.W. Catalytic pyrolysis of lignin over HZSM-5 catalysts: Effect of various parameters on the production of aromatic hydrocarbon. *J. Anal. Appl. Pyrolysis* **2015**, *114*, 273–280. [CrossRef]

17. Overend, R.P.; Chornet, E. Fractionation of lignocellulosics by steam-aqueous pretreatment. *Philos Trans R Soc Lond A* **1987**, *321*, 523–536. [CrossRef]

18. Haghighi Mood, S.; Hossein Golfeshan, A.; Tabatabaei, M.; Salehi Jouzani, G.; Najafi, G.H.; Gholami, M.; Ardjmand, M. Lignocellulosic biomass to bioethanol, a comprehensive review with a focus on pretreatment. *Renew. Sust. Energ. Rev.* **2013**, *27*, 77–93. [CrossRef]

19. Kapoor, M.; Raj, T.; Vijayaraj, M.; Chopra, A.; Gupta, R.P.; Tuli, D.K.; Kumar, R. Structural features of dilute acid, steam exploded, and alkali pretreated mustard stalk and their impact on enzymatic hydrolysis. *Carbohydr. Polym.* **2015**, *124*, 265–273. [CrossRef] [PubMed]

20. Chen, W.H.; Pen, B.L.; Yu, C.T.; Hwang, W.S. Pretreatment efficiency and structural characterization of rice straw by an integrated process of dilute-acid and steam explosion for bioethanol production. *Bioresour. Technol.* **2011**, *102*, 2916–2924. [CrossRef] [PubMed]

21. Zimbardi, F.; Viola, E.; Nanna, F.; Larocca, E.; Cardinale, M.; Barisano, D. Acid impregnation and steam explosion of corn stover in batch processes. *Ind. Crops Prod.* **2007**, *26*, 195–206. [CrossRef]

22. Fockink, D.H.; Sánchez, J.H.; Ramos, L.P. Comprehensive analysis of sugarcane bagasse steam explosion using autocatalysis and dilute acid hydrolysis (H_3PO_4 and H_2SO_4) at equivalent combined severity factors. *Ind. Crops Prod.* **2018**, *123*, 563–572. [CrossRef]

23. Park, S.; Baker, J.O.; Himmel, M.E.; Parilla, P.A.; Johnson, D.K. Cellulose crystallinity index: measurement techniques and their impact on interpreting cellulase performance. *Biotechnol. biofuels* **2010**, *3*, 10. [CrossRef] [PubMed]

Article

A Simultaneous Conversion and Extraction of Furfural from Pentose in Dilute Acid Hydrolysate of *Quercus mongolica* Using an Aqueous Biphasic System

Jong-Hwa Kim [1], Seong-Min Cho [1], June-Ho Choi [1], Hanseob Jeong [2], Soo Min Lee [2], Bonwook Koo [3,*] and In-Gyu Choi [4,5,*]

[1] Department of Forest Sciences College of Agriculture and Life Sciences, Seoul National University, Seoul 08826, Korea; wmfty@snu.ac.kr (J.-H.K.); csmin93@snu.ac.kr (S.-M.C.); jhchoi1990@snu.ac.kr (J.-H.C.)
[2] Wood Chemistry Division, Forest Products Department, National Institute of Forest Science, Seoul 02455, Korea; hsj17@korea.kr (H.J.); lesoomin@korea.kr (S.M.L.)
[3] Green and Sustainable Materials R & D Department, Korea Institute of Industrial Technology, Cheonan-si 31056, Korea
[4] Department of Agriculture, Forestry, and Bioresources, College of Agriculture and Life Sciences, Seoul National University, Seoul 08826, Korea
[5] Research Institute of Agriculture and Life Sciences, College of Agriculture and Life Sciences, Seoul National University, Seoul 08826, Korea
* Correspondence: bkoo@kitech.re.kr (B.K.); cingyu@snu.ac.kr (I.-G.C.)

Abstract: This study optimizes furfural production from pentose released in the liquid hydrolysate of hardwood using an aqueous biphasic system. Dilute acid pretreatment with 4% sulfuric acid was conducted to extract pentose from liquid *Quercus mongolica* hydrolysate. To produce furfural from xylose, a xylose standard solution with the same acid concentration of the liquid hydrolysate and extracting solvent (tetrahydrofuran) were applied to the aqueous biphasic system. A response surface methodology was adopted to optimize furfural production in the aqueous biphasic system. A maximum furfural yield of 72.39% was achieved at optimal conditions as per the RSM; a reaction temperature of 170 °C, reaction time of 120 min, and a xylose concentration of 10 g/L. Tetrahydrofuran, toluene, and dimethyl sulfoxide were evaluated to understand the effects of the solvent on furfural production. Tetrahydrofuran generated the highest furfural yield, while DMSO gave the lowest yield. A furfural yield of 68.20% from pentose was achieved in the liquid hydrolysate of *Quercus mongolica* under optimal conditions using tetrahydrofuran as the extracting solvent. The aqueous and tetrahydrofuran fractions were separated from the aqueous biphasic solvent by salting out using sodium chloride, and 94.63% of the furfural produced was drawn out through two extractions using tetrahydrofuran.

Keywords: aqueous biphasic system; dilute acid hydrolysate; furfural production; solvent extraction; response surface methodology

Citation: Kim, J.-H.; Cho, S.-M.; Choi, J.-H.; Jeong, H.; Lee, S.M.; Koo, B.; Choi, I.-G. A Simultaneous Conversion and Extraction of Furfural from Pentose in Dilute Acid Hydrolysate of *Quercus mongolica* Using an Aqueous Biphasic System. *Appl. Sci.* **2021**, *11*, 163. https://dx.doi.org/10.3390/app11010163

Received: 29 October 2020
Accepted: 23 December 2020
Published: 26 December 2020

Publisher's Note: MDPI stays neutral with regard to jurisdictional claims in published maps and institutional affiliations.

1. Introduction

Lignocellulosic biomass is considered as an alternative energy resource that can mitigate the climate change associated with the excessive use of fossil fuels [1]. Owing to its abundance and non-edibility, cellulose in particular, the key component of biomass, is a rich source of carbohydrates and has been applied to value-add to chemicals or biofuels [2]. Hemicellulose and lignin, the other main components of biomass, combine complex and dense forms of cellulose [3]. This physical barrier makes lignocellulosic biomass chemically and microbiologically resistant [4]. This recalcitrance of biomass requires a pretreatment process to ensure that lignocellulosic biomass is utilized properly and efficiently. The purpose of the pretreatment process is to cleave lignin and hemicellulose and obtain cellulose to improve accessibility for chemicals or enzymes [5]. Among several pretreatment methods, dilute acid pretreatment is considered to be a leading pretreatment technology. This is because it is able to enhance the total sugar yield of the process by solubilizing and converting hemicellulose to pentose [6]. As the pretreatment of lignocellulosic biomass is highly demanding in terms of energy and additional processes compared to edible biomass, its economic feasibility is reduced [7]. As such, utilization of all the main components of biomass to generate valuable products is essential to make this resource more economically feasible.

Hemicellulose is a heteropolysaccharide composed of various monosaccharides such as xylose, arabinose, and mannose [8]. Traditionally, it is considered a by-product in the pulping industry, whereby most hemicellulose is dissolved into black liquor and used as a heating source [9]. However, pentose in hemicellulose may also be converted to valuable chemicals in the same manner as cellulose. This will subsequently improve the economic feasibility of the biorefinery industry through the utilization of lignocellulosic biomass.

Furfural is a building block chemical applied to various fields such as fuel, chemicals, polymers, and pharmaceuticals. Most furfural is produced by acid-catalyzed dehydration of pentose derived from hemicellulose [10]. It may also be produced from lignocellulosic biomass directly via acid catalyst treatment. Although this is a simple and mature process, it is characterized by several disadvantages including low furfural yield, generation of undesirable by-products, and difficulties in utilizing the remaining biomass, such as cellulose or lignin [11]. For this reason, many studies have proposed a two-step process in which hemicellulose is hydrolyzed into pentose or pentose-derived oligomers, and then the pentose or the oligomer is dehydrated into furfural in different reaction systems. This separated system offers several advantages including being able to produce a high furfural yield via the optimization of the reaction system for furfural production (catalyst, solvent, reaction condition, etc.) [12]. The separation of furfural from reaction media continues to be a challenge due to several complicated processes [13].

Furfural may be degraded and condensed in the presence of an acidic catalyst and water [14]. In an aqueous furfural production system that uses water as a solvent, the degradation and self-condensation of furfural limits high yields of furfural. This is despite the fact that water is a commonly used solvent for furfural production because it is inexpensive and eco-friendly. To solve this problem, furfural must be separated from the aqueous solvent system immediately after conversion from pentose.

An aqueous biphasic system may be used to separate furfural from water in the system. The aqueous phase contains water, and the acid catalyst converts pentose to furfural, while the organic phase is composed of the organic solvent that absorbs furfural converted into the organic phase. Dichloromethane (DCM) [15], methylisobuthylketone (MIBK) [16], and tetrahydrofuran (THF) [17] have been used as organic solvents for the extraction. These organic solvents are advantageous in terms of conducting a simultaneous process, including the conversion of pentose to furfural and the extraction of furfural. Furfural is produced by an acid catalyst in the aqueous phase and is extracted immediately by the organic phase [18]. Organic solvents may minimize furfural degradation, improving furfural yield [19].

This study aims to optimize furfural production from the dilute acid hydrolysate of *Quercus mongolica*. A xylose standard solution with the same acid concentration of dilute acid hydrolysate and extracting solvent (THF) were applied to the aqueous biphasic system to determine the optimal conditions for furfural production. A response surface methodology (RSM) in which the independent variables were reaction temperature, time, and xylose concentration was adopted to optimize the aqueous biphasic system. The organic solvent was also evaluated to select the solvent that prevented the degradation and self-condensation of furfural.

2. Materials and Methods

2.1. Materials

The xylose standard was purchased from Sigma-Aldrich Korea Co. (Yongin, Republic of Korea). *Quercus mongolica*, a xylan-rich hardwood was used as the raw material for pentose and it was supplied by the National Institute of Forest Science (NIFoS, Seoul, Republic of Korea). The particle size of the raw material was reduced to less than 0.5 mm through grinding and sieving using a sawdust producer and air classifier mill, respectively. The moisture content was less than 5%, and the chemical composition was determined using the Laboratory Analytical Procedure of National Renewable Energy Laboratory (NREL, Golden, CO, USA) [20].

2.2. Dilute Acid Pretreatment for Pentose Production from Lignocellulosic Biomass

Dilute acid pretreatment of *Quercus mongolica* for pentose production was conducted following the methodology described in previous research [21]. Briefly, raw material was mixed with sulfuric acid solution (4%, *w/w*) in an Erlenmeyer flask; the solid to liquid ratio was 1 to 7. Then, the flask was placed in an autoclave (MLS-3020, Sanyo, Osaka, Japan) at 121 °C for 102.3 min; these are the optimal conditions for pentose production from *Quercus mongolica* as described in the previous research [21]. Following the dilute acid pretreatment, the flask was immediately cooled to room temperature in the ice chamber to stop the reaction. Then, the pentose-rich hydrolysate was separated from the solid residue using a Büchner funnel equipped with filter paper (No. 52, Hyundai Micro Co., Seoul, Korea). The chemical composition of the hydrolysate is shown in Table 1.

Table 1. Chemical composition of dilute acid hydrolysate of *Quercus mongolica*.

Component	Concentration (g/L)
Sugars	
Glucose	2.01 ± 0.05
Xylose+mannose+galactose (XMG)	20.02 ± 0.46
Arabinose	1.64 ± 0.06
Sugar derivatives	
Acetic acid	6.17 ± 0.13
Formic acid	0.03 ± 0.00
Furfural	0.32 ± 0.04
5-Hydroxymethylfurfural (5-HMF)	0.01 ± 0.00

2.3. Response Surface Methodology for Optimization of Furfural Production from Xylose Standard

To maximize furfural production from the hydrolysate, the optimum conditions needed to be determined; optimization was conducted using a xylose standard solution. The xylose standard solution was adopted to investigate the relationship between reaction conditions and furfural yield, excluding the effect of impurities. Briefly, 10 mL of xylose standard solution containing a certain concentration of xylose was mixed with 20 mL of organic solvent in a Teflon-lined reactor. The sulfuric acid concentration of the xylose standard solution was adjusted to 4% (*w/w*); this was the same as the dilute acid hydrolysate used to obtain the optimum reaction conditions for the hydrolysate. The reactor was then sealed and soaked in an oil bath that had been pre-heated to a target temperature. The mixed hydrolysate was stirred during the reaction using a magnetic stirrer, and the temperature was maintained for a certain reaction time at the target temperature. After the reaction, the reactor was removed from the oil bath and immediately stored in the ice chamber to cool to room temperature and prevent undesirable reactions.

An RSM was adopted to optimize furfural production using a xylose standard solution. The analysis was conducted based on a 2^3 factorial central design (CCD) using Design Expert 11.1.0.1 software (Stat-Ease, Inc., Minneapolis, MN, USA). Table 2 presents 17 sets of reaction conditions composed of six axial points and a duplication of the central point. The reaction temperature (X1, °C), reaction time (X2, min), and xylose concentration (X3, g/L) were designated as independent variables, while furfural yield (Y1, %) was the dependent variable.

Table 2. 2^3 factorial experimental design varying on 3 factors and results of furfural yield.

No.	Independent Variables			Dependent Variable
	Reaction Temperature $(X_1, °C)$	Reaction Time (X_2, min)	Xylose Concentration $(X_3, g/L)$	Furfural Yield $(Y_1, \%)$
1	140	60	10	6.96
2	180	60	10	69.87
3	140	180	10	39.47
4	180	180	10	62.48
5	140	60	30	6.67
6	180	60	30	66.83
7	140	180	30	46.59
8	180	180	30	59.23
9	126.36	120	20	4.69
10	193.64	120	20	61.33
11	160	19.09	20	0.71
12	160	220.91	20	68.12
13	160	120	3.18	67.99
14	160	120	36.82	67.77
15	160	120	20	67.75
16	160	120	20	67.75
17	160	120	20	68.02

The coded level of the CCD from each run was applied to real independent variables as follows:

Variable = value of central point/variation of coded level per one point

Reaction temperature (°C) = 160/20, reaction time (min) = 120/60, xylose concentration (g/L) = 20/10

Following the optimization of reaction conditions based on the RSM, the extraction solvent was altered from THF to dimethyl sulfoxide (DMSO) or toluene in the optimum conditions to analyze the effects of the extraction solvent.

2.4. Furfural Production from Pentose Derived from Quercus Mongolica

Simultaneous reactions occurred involving the release of pentose from xylose and the conversion of pentose to furfural in the same reactor without a separation process; this places it in a good stead for industrial application. The pentose derived from xylose was converted to furfural in the same reactor, used for pentose release. The dilute acid hydrolysate of *Quercus mongolica* was mixed with 4% sulfuric acid to adjust xylose concentration to optimal conditions in the Teflon-lined reactor; the total volume was adjusted to 10 mL. Then, 20 mL of THF (i.e., the organic solvent), was added into reactor for furfural extraction. The reaction temperature and time were set to the optimal values as per the results from the RSM analysis. After the reaction, the reactor was removed from the oil bath and immediately stored in the ice chamber to cool to room temperature and prevent undesirable reactions. The organic phase of THF, containing furfural, was separated from the aqueous phase through the addition of NaCl. This was done to investigate the separation efficiency of furfural from the final solution, including furfural and other products.

2.5. Analysis of Furfural and Other Products in Mixed Hydrolysate

The content of furfural and other products such as the remaining sugars was determined using high performance liquid chromatography (Ultimate-3000, Thermo Dionex, Waltham, MA, USA) with an Aminex 87H column (eluent: 0.01 N sulfuric acid, oven temperature: 40 °C, flow rate: 0.5 mL/min). Peaks were identified by comparing the retention time of each peak. The concentration of peaks was identified by comparing the standard calibration curve of each chemical. The furfural yield (Equation (1)) and pentose conversion (Equation (2)) were calculated as follows:

$$\text{Furfural yield (\%)} = \text{furfural after reaction (mol)}/\text{pentose before reaction (mol)} \times 100 \tag{1}$$

$$\text{Pentose conversion (\%)} = (\text{pentose before} - \text{after reaction (mol)})/\text{pentose before reaction (mol)} \times 100 \tag{2}$$

3. Results and Discussion

3.1. Pentose Production during Dilute Acid Pretreatment

Quercus mongolica is chemically composed of 46.67% of glucose, 19.14% of xylose, mannose and galactose (XMG), 0.77% of arabinose, 22.56% of acid insoluble lignin (AIL), 3.19% of acid soluble lignin (ASL), 2.06% of extractives, and 0.05% of ash. It is known that the glycosidic bond between hemicellulose-cellulose is cleaved, and the separated hemicellulose is converted to pentose-like XMG dilute acid pretreatment [22]. As shown in Table 1, the main component in dilute acid hydrolysate was XMG, and the main pentose was xylose [21]. Glucose and arabinose derived from arabinoxylan or glucuronoxylan were also detected; however, they were present in very small amounts compared with pentose. In sugar derivatives, the main product was acetic acid derived from the O-acetyl group in hemicellulose. Although furanic compounds such as 5-hydroxymethylfurfural (5-HMF) and furfural were produced by the acidic dehydration of released sugar, there was only a small amount of these compounds owing to the low severity of the dilute acid pretreatment. This pretreatment had not boosted the dehydration of sugar to the furanic [23].

3.2. RSM for Furfural Production from Xylose Standard Solution with Extracting Solvent

Furfural was produced from the xylose standard by acid-catalyzed dehydration, as listed in Table 2. In run #2 with a reaction temperature of 180 °C, over a 60 min duration using a xylose concentration of 10 g/L, the maximum furfural yield was 69.87% with 96.02% xylose conversion. Although the maximum xylose conversion was 99.32% at run #12, the furfural yield under these conditions was 68.12%; this is lower than the yield in run #2. These results indicate that the more severe experimental conditions of run #12 as a result of a longer reaction time induced the degradation of the furfural that had been produced. The lower furfural yield due to furfural degradation under severe experimental conditions was also observed in runs #4, 8, and 10; these runs were also characterized by long reaction times at certain temperatures. It is inferred that these reaction conditions caused furfural degradation despite the presence of the extraction solvent to prevent furfural degradation. However, a previous study has reported that the amount of furfural degradation due to severe experimental conditions when using an extracting solvent is marginal compared to the amount of furfural degradation in conditions with no extracting solvent [24]. The condensation of furfural was suppressed through the addition of the THF; as such, there were no insoluble precipitates detected from furfural condensation in all experimental conditions.

To evaluate the effect of each variable on furfural yield, regression analysis was undertaken using a 2^3 factorial design matrix with corresponding furfural yield (%). The following quadratic equation was generated (Equation (3)), based on the outcomes of the regression analysis:

$$\text{Furfural yield (\%)} = -1122.2746 + 11.8539\,X_1 + 2.4079\,X_2 + 1.0355\,X_3 - 0.0091\,X_1X_2 - 0.0082\,X_1X_3 + 0.0015\,X_2X_3 - 0.0302\,X_1{}^2 - 0.0032\,X_2{}^2 + 0.0024\,X_3{}^2 \tag{3}$$

In the equation, X_1, X_2, and X_3 represent the actual reaction temperature, reaction time, and xylose concentration, respectively. The model had a high regression coefficient ($R^2 = 0.95$), indicating 95% variability in the response, while the *p*-value was extremely low (0.001), indicating that this regression model was significant. The coefficient of variation (CV) was 18.74%, which indicates the high precision and reliability of the experiments [25].

A three-dimensional (3D) plot and detailed contour of the RSM for furfural yield was established using Equation (3), by varying the three variables within the experimental range (Figure 1). As shown in Figure 1a, the furfural yield increased with reaction temperature and time to approximately 185 °C and 180 min, respectively. Once the temperature and time exceeded these points, there was a decrease in furfural yield due to its degradation under severe experimental conditions [26].

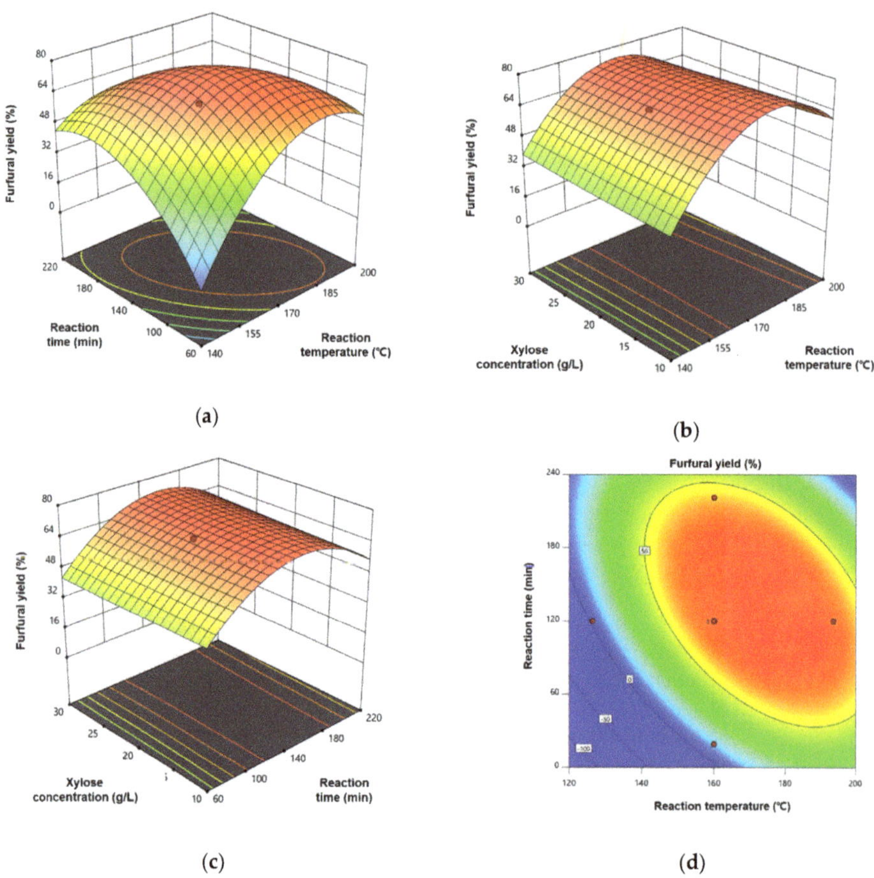

Figure 1. Three-dimensional plot and detailed contour of response surface methodology of furfural yield from xylose standard solution with extraction solvent (THF): (**a**) fixed xylose concentration (10 g/L); (**b**) fixed reaction time (120 min); (**c**) fixed reaction temperature (160 °C); (**d**) contour of furfural yield depending on reaction temperature and time.

This phenomenon was clearly observed in Figure 1b,c. Figure 1b depicts the furfural yield based on the reaction temperature and xylose concentration with a fixed reaction time of 120 min. The furfural yield increased with temperature until approximately 165 °C, and then remained stable at temperatures of 165–185 °C, which was where the maximum yield was observed. Then, yield decreased when the temperature increased above 185 °C regardless of the xylose concentration. These results are similar to the findings from previous research where furfural production from biomass occurred using sulfuric acid as a catalyst [27,28]. Figure 1c depicts the relationship between the furfural yield, the reaction time, and xylose concentration with a fixed reaction temperature of 160 °C. The furfural yield increased with reaction time up to 180 min. Beyond this time, yield decreased due to furfural decomposition [29] and self-condensation [30]. The xylose concentration was considered a less influential variable when compared to reaction temperature and time, as shown in Figure 1b,c. This inference was also supported by the p-value of the variables (Table 3).

The sources related to xylose concentration (X_3, X_1X_3, X_2X_3, X_3^2) had high p-values indicating that xylose concentration was not as significant a factor for furfural yield. This result differs from previous research, which has demonstrated that furfural yield generally has an inverse relationship with xylose concentration. Higher xylose concentrations are able to produce more furfural at once, leading to a higher collision possibility that causes condensation of the furfural reaction [31]. It was assumed that the 10–30 g/L xylose concentration range used in this study was too narrow of a range to affect furfural yield. This was unlike a study by Yang [32] where they produced furfural at high xylose concentrations. Yang [32] varied the xylose concentration from 40 to 120 g/L and found that furfural yield had been relatively stable when the xylose concentration was from 40 to 70 g/L. Based on the xylose concentration range used in this study (10–30 g/L), the effect on furfural yield was negligible compared to reaction time and temperature. With a fixed xylose concentration there was greater clarity in terms of the optimal conditions to maximize furfural yield. Additionally, the reaction temperature and time were expanded compared to the experimental range (Figure 1d).

Table 3. Analysis of variance (ANOVA) for furfural yield in dehydration of xylose model compound and coefficients for quadratic equation.

Source	Sum of Squares	DF	Mean Square	F-Value	p-Value	Coefficient
Model	10,697.21	9	1188.58	14.13	0.0010	67.77
X_1	4722.88	1	4722.88	56.14	0.0001	18.60
X_2	2136.23	1	2136.23	25.39	0.0015	12.51
X_3	0.00	1	0.00	0.00	0.9963	0.01
X_1X_2	955.35	1	955.35	11.36	0.0119	−10.93
X_1X_3	21.51	1	21.51	0.26	0.6287	−1.64
X_1X_3	6.48	1	6.48	0.08	0.7894	0.90
X_1^2	1646.23	1	1646.23	19.57	0.0031	−12.08
X_2^2	1513.50	1	1513.50	17.99	0.0038	−11.59
X_3^2	0.67	1	0.67	0.01	0.9314	0.24
Residual	588.94	7	84.13			
Lack of fit	588.89	5	117.78	4825.36	0.0002	
Pure error	0.05	2	0.02			
Corrected total	11,286.15	16				

The optimal conditions to maximize furfural yield was calculated based on Equation (3). The maximum furfural yield in the predicted reaction conditions was 75.1%, where the reaction temperature was 170 °C, reaction time was 120 min, and xylose concentration was 10 g/L. To verify the model, actual furfural production was carried out using these predicted optimal conditions, rendering a furfural yield of 72.39%, similar to the predicted yield.

3.3. Effect of Organic Solvent for Furfural Production and Extraction

Three kinds of organic solvents, THF, toluene, and DMSO, were evaluated to understand the effects of organic solvents on furfural production and extraction. Toluene is considered an effective solvent for furfural extraction [33], whereby it does not require additional salt for phase separation because of its immiscibility with water. DMSO has been used to improve the selectivity of 5-HMF from glucose by increasing the fructofuranose isomer and stabilizing 5-HMF by hydrogen bonding [34]. Under a similar mechanism, it was anticipated that DMSO could also improve furfural yield from xylose. Reaction conditions were set to the optimal values predicted from analysis of variance (ANOVA); this was a reaction temperature of 170 °C, reaction time of 120 min, xylose concentration of 10 g/L of, and the use of a 4% of sulfuric acid solution.

Table 4 shows the furfural yield from the xylose standard solution depending on the organic solvent. THF had the highest furfural yield from the xylose standard, while DMSO had the lowest among the three solvents. It was assumed that DMSO may not effectively protect the generated furfural from acid or water as furfural has no hydroxyl group compared with 5-HMF. In addition, DMSO has a reduced interaction with furfural compared to the other organic solvents such as toluene or THF, as its polarity is higher than that of furfural. THF had a higher furfural yield than toluene, even though the polarity of toluene was close to furfural. To explain this phenomenon, the partition coefficient of furfural in THF/water and toluene/water was calculated by dividing furfural concentration in the organic solvent by the concentration of furfural in aqueous water (Table 4) to compare the solubility of furfural in a two-liquid mixture (water and organic solvent). To calculate the partition coefficient of furfural to THF/water, NaCl was added to separate THF from the water phase (salting out). THF had a higher partition coefficient than toluene, indicating that the former was more effective in extracting furfural than toluene, and yielding higher amounts of furfural.

Table 4. Properties of organic solvent and furfural yield from xylose standard based on the type of organic solvent utilized.

Solvent	THF	Toluene	DMSO
Spectroscopic polarity (Furfural: 0.426 [35])	0.6 [36]	0.55 [37]	1 [38]
Partition coefficient *	9.05	5.82	N/D **
Furfural yield (%)	72.39 ± 0.50	58.01 ± 0.00	38.28 ± 0.00

* Partition coefficient = [Furfural]org/[Furfural]aq, ** N/D: Not detected.

3.4. Production of Furfural from Pentose in Dilute Acid Hydrolysate

The optimal reaction conditions as predicted using the ANOVA (i.e., 170 °C, 120 min, and 10 g/L of xylose) were adopted to maximize furfural production from pentose in dilute acid hydrolysate. The xylose concentration was adjusted by mixing the hydrolysate, which had already contained 4% (w/w) sulfuric acid, with 4% sulfuric acid solution. The total volume of the aqueous phase was adjusted to 10 mL, similar to the xylose standard solution, and 20 mL of THF was added to extract furfural from the aqueous phase.

Table 5 describes the pentose conversion and furfural yield from the xylose standard and dilute acid hydrolysate. The furfural yield and pentose conversion of hydrolysate were slightly lower, compared with the xylose standard. The presence of impurities such as hexoses, organic acids, and acid soluble lignin, may have impacted on the extraction efficiency of furfural from the hydrolysate [39].

Table 5. Pentose conversion and furfural yield from xylose standard and dilute acid hydrolysate.

	Pentose Conversion (%)	Furfural Yield (%)
Xylose standard solution	100 ± 0.00	72.39 ± 0.50
Liquid hydrolysate	94.69 ± 0.76	68.20 ± 0.20

The difference in furfural yield between the xylose standard and the hydrolysate was not as considerable as had been expected. This indicates that THF effectively prevents furfural loss by inhibiting the ring opening of furfural and condensation between furfural and acid soluble lignin to form insoluble precipitate [40].

To investigate the distribution of impurities in the organic and aqueous phases, NaCl was added to separate the hydrolysate into the organic and aqueous phases. Table 6 presents the change in the distribution of furfural and other chemicals in furfural that produced hydrolysate by phase separation. Most furfural produced was extracted in the organic phase with a partition coefficient of 8.43; this is slightly lower than that of the xylose standard solution (9.05). Sugars, such as glucose, XMG, and arabinose favor the aqueous phase owing to their hydroxyl group, while other chemicals such as furfural, organic acids, and ASL tend to be extracted by the organic phase of THF. It is known that THF is able to effectively dissolve lignin as it has a high affinity to phenolic compounds [41], thus, most ASL was extracted to THF. Approximately three-quarters of organic acids were extracted to the organic phase, and the THF was effective in extracting various organic acids; however, the detailed extraction mechanism of THF to organic acid continues to be unclear [42,43].

Table 6. Concentration of chemicals in furfural produced liquor prior to and after phase separation (organic phase, aqueous phase) through the addition of NaCl.

	Concentration (g/L)						
	Furfural	Glucose	XMG	Arabinose	Formic Acid	Acetic Acid	Acid Soluble Lignin
Before separation	3.08 ± 0.11	0.09 ± 0.02	0.22 ± 0.03	0.05 ± 0.02	0.49 ± 0.02	1.22 ± 0.03	0.84 ± 0.02
Organic phase	4.84 ± 0.10	0.00 ± 0.00	0.09 ± 0.00	0.00 ± 0.00	0.63 ± 0.02	1.61 ± 0.04	1.24 ± 0.00
Aqueous phase	0.58 ± 0.05	0.25 ± 0.05	0.29 ± 0.05	0.05 ± 0.00	0.21 ± 0.00	0.55 ± 0.03	0.21 ± 0.00

THF, the separated organic phase, was removed, and 20 mL of fresh THF was added to 10 mL of furfural extracted aqueous phase to enhance the extraction rate of furfural to the organic phase. After each additional extraction, the amount of extracted furfural and other compounds were analyzed; the extraction rate was calculated as (Equation (4)):

$$\text{Extraction rate (\%)} = \text{Amount of products extracted in THF phase (g)} / \text{Amount of products existed in furfural produced liquor before phase separation (g)} \tag{4}$$

Figure 2 shows the increase in the furfural extraction rate to THF with the number of extractions. Following the second additional extraction, the extraction rate increased from 86.03% to 94.63%. However, impurities such as organic acids and ASL had also been further extracted as the number of extractions increased. In particular, organic acids such as formic acid and acetic acid were completely extracted to THF following the second extraction. This means that additional impurity separation is required to acquire furfural with higher purity. It was reported that organic acids and ASL may be separated from THF by the ion exchange resin [44,45], and absorbents such as activated carbon, respectively [46]. However, lignin separation by an absorbent must occur prior to furfural production, as the absorbent absorbs furfural and ASL through a π-π interaction [47].

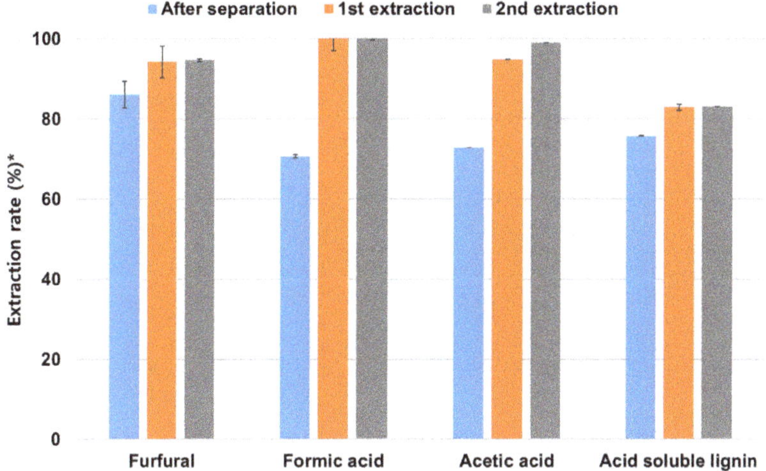

Figure 2. Extraction rate of products in furfural generated hydrolysate by extracting solvent (THF).

4. Conclusions

This study aimed to optimize furfural production from pentose in the dilute acid hydrolysate of *Quercus mongolica*. The main component of the acid hydrolysate was XMG, which was dominated by xylose. The optimization of furfural production was conducted using RSM with a xylose standard solution and an extracting solvent THF, to enhance furfural yield. The optimal conditions included a reaction temperature of 170 °C at a reaction time of 120 min with a xylose concentration of 10 g/L; the predicted furfural yield was 75.1% under these conditions. An experimental furfural yield of 72.39% was obtained under the optimal experimental conditions, similar to the predicted yield. Extracted solvents such as THF, toluene, and DMSO were evaluated to understand the effect of the solvent on furfural yield. THF achieved the highest furfural yield, while DMSO had the lowest yield. Based on this result, furfural was produced from dilute acid hydrolysate under optimized reaction conditions using THF as the extracting solvent. A furfural yield of 68.20% was obtained based on pentose in the hydrolysate, similar to that of the xylose standard solution (72.39%). Following phase separation through the addition of NaCl 86.03% of the furfural produced was in the organic phase. The THF, and two additional extractions using fresh THF enhanced the furfural extraction rate to 94.63%.

Author Contributions: Conceptualization, J.-H.K. and B.K.; methodology, S.-M.C. and J.-H.K formal analysis, J.-H.K. and J.-H.C.; investigation, J.-H.K.; resources, H.J. and S.M.L.; data curation J.-H.K. and J.-H.C.; writing-original draft preparation, J.-H.K.; writing-review and editing, J.-H.K visualization, J.-H.K.; supervision, I.-G.C. and B.K.; project administration, B.K. All authors have read and agreed to the published version of the manuscript.

Funding: This study was carried out with the support of the R & D program for Forest Science Technology (Project No. 2020226C10-2022-AC01) provided by the Korea Forest Service (Korea Forestry Promotion Institute) and the Research Program (FP0900-2019-01) of the National Institute of Forest Science (NIFoS) (Seoul, Republic of Korea).

Institutional Review Board Statement: Not applicable.

Informed Consent Statement: Not applicable.

Conflicts of Interest: The authors declare no conflict of interest.

References

1. Stöcker, M. Biofuels and biomass-to-liquid fuels in the biorefinery: Catalytic conversion of lignocellulosic biomass using porous materials. *Angew. Chem. Int. Ed.* **2008**, *47*, 9200–9211. [CrossRef] [PubMed]
2. Zhou, C.-H.; Xia, X.; Lin, C.-X.; Tong, D.-S.; Beltramini, J. Catalytic conversion of lignocellulosic biomass to fine chemicals and fuels. *Chem. Soc. Rev.* **2011**, *40*, 5588–5617. [CrossRef] [PubMed]
3. Liu, Y.; Chen, W.; Xia, Q.; Guo, B.; Wang, Q.; Liu, S.; Liu, Y.; Li, J.; Yu, H. Efficient cleavage of lignin-carbohydrate complexes and ultrafast extraction of lignin oligomers from wood biomass by microwave-assisted treatment with deep eutectic solvent *ChemSusChem* **2017**, *10*, 1692–1700. [CrossRef] [PubMed]
4. Zhao, X.; Zhang, L.; Liu, D. Biomass recalcitrance. Part I: The chemical compositions and physical structures affecting the enzymatic hydrolysis of lignocellulose. *Biofuels Bioprod. Biorefining* **2012**, *6*, 465–482. [CrossRef]
5. Mosier, N.; Wyman, C.; Dale, B.; Elander, R.; Lee, Y.; Holtzapple, M.; Ladisch, M.R. Features of promising technologies for pretreatment of lignocellulosic biomass. *Bioresour. Technol.* **2005**, *96*, 673–686. [CrossRef] [PubMed]
6. Wyman, C.E.; Dale, B.E.; Elander, R.T.; Holtzapple, M.; Ladisch, M.R.; Lee, Y. Coordinated development of leading biomass pretreatment technologies. *Bioresour. Technol.* **2005**, *96*, 1959–1966. [CrossRef]
7. Zhu, J.Y.; Pan, X.; Zalesny, R.S., Jr. Pretreatment of woody biomass for biofuel production: Energy efficiency, technologies, and recalcitrance. *Appl. Microbiol. Biotechnol.* **2010**, *87*, 847–857. [CrossRef]
8. Puls, J. Chemistry and biochemistry of hemicelluloses: Relationship between hemicellulose structure and enzymes required for hydrolysis. *Macromol. Symp.* **1997**, *120*, 183–196. [CrossRef]
9. Farhat, W.; Venditti, R.A.; Hubbe, M.; Taha, M.; Becquart, F.; Ayoub, A. A Review of water-resistant hemicellulose-based materials Processing and applications. *ChemSusChem* **2017**, *10*, 305–323. [CrossRef]
10. Mansilla, H.D.; Baeza, J.; Urzua, S.; Maturana, G.; Villaseñor, J.; Durán, N. Acid-catalysed hydrolysis of rice hull: Evaluation of furfural production. *Bioresour. Technol.* **1998**, *66*, 189–193. [CrossRef]
11. Rivas, S.; Vila, C.; Santos, V.; Parajó, J.C. Furfural production from birch hemicelluloses by two-step processing: A potential technology for biorefineries. *Holzforschung* **2016**, *70*, 901–910. [CrossRef]
12. Wang, X.; Zhang, C.; Lin, Q.; Cheng, B.; Kong, F.; Li, H.; Ren, J. Solid acid-induced hydrothermal treatment of bagasse for production of furfural and levulinic acid by a two-step process. *Ind. Crop. Prod.* **2018**, *123*, 118–127. [CrossRef]
13. Weingarten, R.; Cho, J.; Conner, W.C., Jr.; Huber, G.W. Kinetics of furfural production by dehydration of xylose in a biphasic reactor with microwave heating. *Green Chem.* **2010**, *12*, 1423–1429. [CrossRef]
14. Hu, X.; Westerhof, R.J.M.; Dong, D.; Wu, L.; Li, C.-Z. Acid-catalyzed conversion of xylose in 20 solvents: Insight into interactions of the solvents with xylose, furfural, and the acid catalyst. *ACS Sustain. Chem. Eng.* **2014**, *2*, 2562–2575. [CrossRef]
15. Deng, A.; Lin, Q.; Yan, Y.; Li, H.; Ren, J.; Liu, C.; Sun, R. A feasible process for furfural production from the pre-hydrolysis liquor of corncob via biochar catalysts in a new biphasic system. *Bioresour. Technol.* **2016**, *216*, 754–760. [CrossRef]

16. Chen, Z.; Bai, X.; Lusi, A.; Jacoby, W.A.; Wan, C. One-pot selective conversion of lignocellulosic biomass into furfural and co-products using aqueous choline chloride/methyl isobutyl ketone biphasic solvent system. *Bioresour. Technol.* **2019**, *289*, 121708. [CrossRef]

17. Xu, S.; Pan, D.; Wu, Y.; Song, X.; Gao, L.; Li, W.; Das, L.; Xiao, G. Efficient production of furfural from xylose and wheat straw by bifunctional chromium phosphate catalyst in biphasic systems. *Fuel Process. Technol.* **2018**, *175*, 90–96. [CrossRef]

18. Gürbüz, E.I.; Wettstein, S.G.; Dumesic, J.A. Conversion of hemicellulose to furfural and levulinic acid using biphasic reactors with alkylphenol solvents. *ChemSusChem* **2012**, *5*, 383–387. [CrossRef]

19. Li, H.; Deng, A.; Ren, J.; Liu, C.; Wang, W.; Peng, F.; Sun, R.-C. A modified biphasic system for the dehydration of d-xylose into furfural using $SO_{42}-/TiO_2$-ZrO_2/La_{3+} as a solid catalyst. *Catal. Today* **2014**, *234*, 251–256. [CrossRef]

20. Sluiter, A.; Hames, B.; Ruiz, R.; Scarlata, C.; Sluiter, J.; Templeton, D.; Crocker, D. Determination of structural carbohydrates and lignin in biomass. *Lab. Anal. Proced.* **2008**, *1617*, 1–16.

21. Jang, S.-K.; Kim, J.-H.; Choi, I.-G.; Choi, J.-H.; Lee, S.-M.; Choi, I.-G. Investigation of conditions for dilute acid pretreatment for improving xylose solubilization and glucose production by supercritical water hydrolysis from Quercus mongolica. *Renew. Energy* **2018**, *117*, 150–156. [CrossRef]

22. Cabiac, A.; Guillon, E.; Chambon, F.; Pinel, C.; Rataboul, F.; Essayem, N. Cellulose reactivity and glycosidic bond cleavage in aqueous phase by catalytic and non catalytic transformations. *Appl. Catal. A Gen.* **2011**, *402*, 1–10. [CrossRef]

23. Karimi, K.; Kheradmandinia, S.; Taherzadeh, M.J. Conversion of rice straw to sugars by dilute-acid hydrolysis. *Biomass Bioenergy* **2006**, *30*, 247–253. [CrossRef]

24. Li, H.; Ren, J.; Zhong, L.; Sun, R.; Liang, L. Production of furfural from xylose, water-insoluble hemicelluloses and water-soluble fraction of corncob via a tin-loaded montmorillonite solid acid catalyst. *Bioresour. Technol.* **2015**, *176*, 242–248. [CrossRef] [PubMed]

25. Thong, S.O.; Prasertsan, P.; Intrasungkha, N.; Dhamwichukorn, S.; Birkeland, N.-K. Optimization of simultaneous thermophilic fermentative hydrogen production and COD reduction from palm oil mill effluent by Thermoanaerobacterium-rich sludge. *Int. J. Hydrogen Energy* **2008**, *33*, 1221–1231. [CrossRef]

26. Qing, Q.; Guo, Q.; Zhou, L.; Wan, Y.; Xu, Y.; Ji, H.; Gao, X.; Zhang, Y. Catalytic conversion of corncob and corncob pretreatment hydrolysate to furfural in a biphasic system with addition of sodium chloride. *Bioresour. Technol.* **2017**, *226*, 247–254. [CrossRef] [PubMed]

27. Jang, S.-K.; Choi, I.-G.; Hong, C.-Y.; Kim, H.-Y.; Ryu, G.-H.; Yeo, H.; Choi, J.W.; Choi, I.-G. Changes of furfural and levulinic acid yield from small-diameter Quercus mongolica depending on dilute acid pretreatment conditions. *J. Korean Wood Sci. Technol.* **2015**, *43*, 838–850. [CrossRef]

28. López, F.; García, M.; Feria, M.; García, J.; De Diego, C.; Zamudio, M.; Diaz, M. Optimization of furfural production by acid hydrolysis of Eucalyptus globulus in two stages. *Chem. Eng. J.* **2014**, *240*, 195–201. [CrossRef]

29. Zhang, T.; Kumar, R.; Wyman, C.E. Enhanced yields of furfural and other products by simultaneous solvent extraction during thermochemical treatment of cellulosic biomass. *RSC Adv.* **2013**, *3*, 9809–9819. [CrossRef]

30. Shao, Y.; Hu, X.; Zhang, Z.; Sun, K.; Gao, G.; Wei, T.; Zhang, S.; Hu, S.; Xiang, J.; Wang, Y. Direct conversion of furfural to levulinic acid/ester in dimethoxymethane: Understanding the mechanism for polymerization. *Green Energy Environ.* **2019**, *4*, 400–413. [CrossRef]

31. Zhu, Y.; Liab, W.; Lu, Y.; Zhang, T.; Jameel, H.; Chang, H.-M.; Ma, L. Production of furfural from xylose and corn stover catalyzed by a novel porous carbon solid acid in γ-valerolactone. *RSC Adv.* **2017**, *7*, 29916–29924. [CrossRef]

32. Yang, W.; Liac, P.; Bo, D.; Chang, H. The optimization of formic acid hydrolysis of xylose in furfural production. *Carbohydr. Res.* **2012**, *357*, 53–61. [CrossRef] [PubMed]

33. Gupta, N.K.; Fukuoka, A.; Nakajima, K. Amorphous Nb_2O_5 as a selective and reusable catalyst for furfural production from xylose in biphasic water and toluene. *ACS Catal.* **2017**, *7*, 2430–2436. [CrossRef]

34. Whitaker, M.R.; Parulkar, A.; Brunelli, N.A. Selective production of 5-hydroxymethylfurfural from fructose in the presence of an acid-functionalized SBA-15 catalyst modified with a sulfoxide polymer. *Mol. Syst. Des. Eng.* **2020**, *5*, 257–268. [CrossRef]

35. Zeitsch, K.J. *The Chemistry and Technology of Furfural and Its Many By-Products*, 1st ed.; Elsevier: Amsterdam, The Netherlands, 2000; Volume 13.

36. Cheong, W.J.; Carr, P.W. Kamlet-Taft pi* polarizability/dipolarity of mixtures of water with various organic solvents. *Anal. Chem.* **1988**, *60*, 820–826. [CrossRef]

37. Leggett, D.C. Modeling solvent extraction using the solvatochromic parameters .alpha., .beta., and .pi.*. *Anal. Chem.* **1993**, *65*, 2907–2909. [CrossRef]

38. Crowhurst, L.; Falcone, R.; Lancaster, N.L.; Llopis-Mestre, V.; Welton, T. Using kamlet-taft solvent descriptors to explain the reactivity of anionic nucleophiles in ionic liquids. *J. Org. Chem.* **2006**, *71*, 8847–8853. [CrossRef]

39. Zhang, T.; Liab, W.; Xu, Z.; Liu, Q.; Ma, Q.; Jameel, H.; Chang, H.-M.; Ma, L. Catalytic conversion of xylose and corn stalk into furfural over carbon solid acid catalyst in γ-valerolactone. *Bioresour. Technol.* **2016**, *209*, 108–114. [CrossRef]

40. Liu, H.; Hu, H.; Jahan, M.S.; Ni, Y. Furfural formation from the pre-hydrolysis liquor of a hardwood kraft-based dissolving pulp production process. *Bioresour. Technol.* **2013**, *131*, 315–320. [CrossRef]

41. Nguyen, T.Y.; Cai, C.M.; Kumar, R.; Wyman, C.E. Co-solvent pretreatment reduces costly enzyme requirements for high sugar and ethanol yields from lignocellulosic biomass. *ChemSusChem* **2015**, *8*, 1716–1725. [CrossRef]

42. Wittmann, G.; Karg, E.; Mühl, A.; Bodamer, O.A.; Túri, S. Comparison of tetrahydrofuran and ethyl acetate as extraction solvents for urinary organic acid analysis. *J. Inherit. Metab. Dis.* **2008**, *31*, 73–80. [CrossRef] [PubMed]
43. Xing, R.; Qi, W.; Huber, G.W. Production of furfural and carboxylic acids from waste aqueous hemicellulose solutions from the pulp and paper and cellulosic ethanol industries. *Energy Environ. Sci.* **2011**, *4*, 2193–2205. [CrossRef]
44. Ahsan, L.; Jahan, S.; Khan, M.I.H.; Calhoun, L. Recovery of acetic acid from prehydrolysis liquor of kraft hardwood dissolving pulp using ion-exchange resin. *Bioresources* **2014**, *9*, 1588–1595. [CrossRef]
45. Uslu, H. Adsorption equilibria of formic acid by weakly basic adsorbent Amberlite IRA-67: Equilibrium, kinetics, thermodynamic *Chem. Eng. J.* **2009**, *155*, 320–325. [CrossRef]
46. Singh, S.; Verma, S.; Mall, I.D. Fixed-bed study for adsorptive removal of furfural by activated carbon. *Colloids Surf. A Physicochem. Eng. Asp.* **2009**, *332*, 50–56. [CrossRef]
47. Choi, I.-G.; Lee, J.; Ju, Y.M.; Lee, S.M. Using electro-coagulation treatment to remove phenolic compounds and furan derivatives in hydrolysates resulting from pilot-scale supercritical water hydrolysis of Mongolian oak. *Renew. Energy* **2019**, *138*, 971–979. [CrossRef]

applied
sciences

Article

Environmentally Friendly Approach for the Production of Glucose and High-Purity Xylooligosaccharides from Edible Biomass Byproducts

Soo-Kyeong Jang [1], Chan-Duck Jung [2], Ju-Hyun Yu [2] and Hoyong Kim [2,*]

[1] Department of Wood Science, Faculty of Forestry, The University of British Columbia, 2424 Main Mall, Vancouver, BC V6T 1Z4, Canada; sookyeong.jang@ubc.ca
[2] Center for Bio-Based Chemistry, Korea Research Institute of Chemical Technology, Ulsan 44429, Korea; cksejrl@krict.re.kr (C.-D.J.); jhyu@krict.re.kr (J.-H.Y.)
* Correspondence: hykim03@krict.re.kr; Tel.: +82-52-241-6357

Received: 26 October 2020; Accepted: 13 November 2020; Published: 16 November 2020

Abstract: Xylooligosaccharides (XOS) production from sweet sorghum bagasse (SSB) has been barely studied using other edible biomasses. Therefore, we evaluated the XOS content as well as its purity by comparing the content of total sugars from SSB. An environmentally friendly approach involving autohydrolysis was employed, and the reaction temperature and time had variations in order to search for the conditions that would yield high-purity XOS. After autohydrolysis, the remaining solid residues, the glucan-rich fraction, were used as substrates to be enzymatically hydrolyzed for glucose conversion. The highest XOS was observed for total sugars (68.7%) at 190 °C for 5 min among the autohydrolysis conditions. However, we also suggested two alternative conditions, 180 °C for 20 min and 190 °C for 15 min, because the former condition might have the XOS at a low degree of polymerization with a high XOS ratio (67.6%), while the latter condition presented a high glucose to total sugar ratio (91.4%) with a moderate level XOS ratio (64.4%). Although it was challenging to conclude on the autohydrolysis conditions required to obtain the best result of XOS content and purity and glucose yield, this study presented approaches that could maximize the desired product from SSB, and additional processes to reduce these differences in conditions may warrant further research.

Keywords: xylooligosaccharides; autohydrolysis; enzymatic hydrolysis; sweet sorghum bagasse

1. Introduction

Due to an emerging interest in a healthy lifestyle, probiotics have attracted attention [1]. Probiotics have several beneficial effects on the human body, especially as it reduces the risk of colon-related diseases and dysfunctions [2]. Prebiotics have also attracted great interest because they facilitate the growth of probiotics and inhibit the growth of pathogenic microorganisms in the human intestine [3].

Xylooligosaccharides (XOS) are one of the fascinating products of biomass that have unique properties as prebiotics [4]. XOS do not only reduce the risk of colon cancer, but also improve beneficial biological activities, including reducing dental caries, boosting the immune system, and restraining the growth of pathogens [5]. Food and food additives have been recognized as conventional methods for XOS ingestion, but the application area of XOS is expanding the pharmaceutical, chemical, and nutraceutical industries [6]. The global market size of XOS is expected to grow by 4.1% annually and reach 135.7 million dollars by 2026 [7].

XOS is typically defined as two to ten xylose combined by β-1,4 linkages with arabinose, acetyl groups, and uronic acid substitution [8]. Chemical (acid pretreatment and autohydrolysis) and biological (enzymatic hydrolysis by xylanase) approaches have been developed for hydrolyzing and extracting the xylan chain in a biomass [9]. Concentrated acid pretreatment using inorganic acids has been reported to result in fast isolation of the xylan fraction, but this method is not environmentally friendly and corrodes the equipment [10]. Meanwhile, the biological method can control the generation of byproducts, as shown in the results of acid pretreatment, but it has drawbacks such as long reaction time, relatively low XOS yield, and high cost of xylanase [11,12]. Autohydrolysis is considered an eco-friendly pretreatment method because it does not use any chemicals for the reaction [13]. When a biomass is heated in water, the uronic and acetyl groups from the hemicellulose fraction can be released in the form of an acid followed by acid-hydrolysis of the carbohydrate [14]. This acidic circumstance of autohydrolysis from the side groups of hemicellulose does not lead to an extremely low pH condition compared to the mineral acids of typical acid pretreatments [15]. Thus, autohydrolysis can be a less corrosive and low-cost approach for XOS production than other chemical methods [13].

Several types of xylan-rich biomasses are used for producing XOS, such as corn stover, wheat straw, sugarcane straw, and sugarcane bagasse [16–19]. However, the straws from agricultural residues usually contain a high amount of minerals, including amorphous silica, which can cause chemical reactions to occur in an undesired manner [20]. The production of numerous XOS has been focused on the utilization of sugarcane bagasse, while interest for sweet sorghum is still lower than that of other biomasses [21]. Sweet sorghum can be cultivated in most regions of the world, from temperate to tropical [22]. It typically grows over 3 m tall, ensuring a high energy density per cultivating area up to 20.2 tons per ha [23]. Additionally, sweet sorghum bagasse (SSB) contains an abundant amount of xylan, which is an advantage for XOS production [24].

The physiological activities of XOS have been considered to be strongly related to the linkage-type, substituted side groups, and the degree of polymerization (DP) [25]. Regarding the DP, XOS has a low DP from xylobiose (DP = 2) to xylotetraose (DP = 4), and performs better as a prebiotic than XOS with a high DP [26,27]. On the other hand, xylose and furfural, which are dehydrated products from xylose, can be excessively produced when high temperature or/and low pH conditions are induced to focus on the cleavage of the xylan structure [28]. Therefore, suitable autohydrolysis conditions for improving the XOS yield and purity should be investigated. In addition, when the high-valued XOS containing a dominant amount of DP 2-4 XOS produced with additional xylanase treatment as a follow-up step, the optimum conditions can reduce the purification and production costs. In this study, SSB, a promising biomass, was subjected to the autohydrolysis process to produce XOS. The conditions were evaluated for a high ratio of XOS to total sugars. After autohydrolysis, the remaining solid residues, cellulose-rich fraction, were used for enzymatic hydrolysis to produce a glucose solution.

2. Materials and Methods

2.1. Feedstock

Sweet sorghum (*Sorghum bicolor* (L.)) bagasse was generously supplied by Good Farmer Co., Ltd., Yecheon-gun, Gyeongsangbuk-do, Korea. The feedstock was oven dried and milled to <0.5 mm using a knife mill. The chemical composition of sweet sorghum is shown in Table 1.

Table 1. Chemical composition of sweet sorghum (g/100 g oven dry weight).

Composition [1]	Sweet Sorghum Bagasse (SSB)
Extractives	9.3 ± 0.4
Carbohydrate	67.3
Glucan	37.6 ± 0.4
XGM [2]	21.2 ± 0.5
Arabinan	2.3 ± 0.0
Acetate	6.1 ± 0.0

Table 1. *Cont.*

Composition [1]	Sweet Sorghum Bagasse (SSB)
Lignin	17.2
Acid-insoluble	15.1 ± 0.1
Acid-soluble	2.1 ± 0.1
Ash	1.1 ± 0.0
Crude protein	6.5 ± 0.0
Total	101.3

[1] All the numbers are based on the initial (OD) weight of the analyzed sample. [2] Sum of xylan, galactan, and mannan.

2.2. Autohydrolysis

Autohydrolysis was conducted to identify a good point to maximize the XOS amount in the liquid hydrolysate. It was carried out using a bomb-type mini-reactor (30 mm ID × 140 mm L, wall thickness 5 mm, total volume 100 mL, KRICT, Daejeon, Korea), equipped with a K-type thermocouple for monitoring the actual reaction temperature. The reaction temperature was set as a variable from 160 to 200 °C at 10 °C intervals. One gram of the oven-dried weight of sweet sorghum bagasse was placed in a custom-made reactor with 20 mL of water. When the reactor reached the designated reaction temperature in an oil bath, it was held for 20 min (reaction time). The reaction time was also changed for 180 and 190 °C conditions from 5 to 20 min with 5 min intervals. After autohydrolysis, slurries were recovered from the reactor into a 50 mL conical tube using an additional 20 mL of water and separated into a solid residue and liquid hydrolysate by centrifugation.

2.3. Enzymatic Hydrolysis

Enzymatic hydrolysis was performed using the solid residue from autohydrolysis. The solid residue obtained from each autohydrolysis condition was directly used as a substrate for the commercial cellulase complex Cellic CTec3 (Novozyme Korea, Seoul, Korea). The dosage of the cellulase complex was 10 FPU/g glucan in the substrate. The citrate buffer solution was adjusted to pH 5.5, and the mixture was incubated in a shaker at 50 °C for 72 h at 150 rpm.

The glucose to total sugar ratio was calculated as follow:

$$\text{Glucose to total sugar ratio (\%)} = \frac{\text{Glucose in the solution after enzymatic hydrolysis (g)}}{\text{Total sugars in the solution after enzymatic hydrolysis (g)}} \times 100 \tag{1}$$

2.4. Determination of the Chemical Composition

The chemical composition of the liquid hydrolysate, enzymatic hydrolysis solution, and the raw material (SSB) were determined using the Laboratory Analytical Procedure of NREL [29,30]. Furthermore, additional acid hydrolysis using 4% sulfuric acid solution was conducted to convert all kinds of carbohydrates in the liquid hydrolysate to monomeric sugars. The amount of monomeric sugars was quantified by high-performance liquid chromatography (HPLC) (Agilent 1200 Infinity, Agilent Technologies Korea Inc., Seoul, Korea). HPLC was equipped with an HPX-87H (300 × 7.8 mm) column (Bio-Rad, Hercules, CA, USA) and a refractive index detector (RID-6A, Bio-Rad, Hercules, CA, USA). Twenty microliters of sample was injected for the HPLC analysis, and the oven temperature was set at 50 °C. The mobile phase was 5 mM sulfuric acid solution and eluted at a 0.6 mL/min flow rate. For quantification, the calibration curves were made using standard materials (glucose, xylose, and arabinose), which were purchased from Sigma-Aldrich Korea Co. (Yongin, Korea). The nitrogen content of SSB was determined with an elemental analyzer (Flash EA 2000, Thermo Electron Corporation, Waltham, MA, USA), and crude protein content was calculated using nitrogen content and conversion factor (6.25) [31].

The XOS to total sugar ratio was calculated by following formula:

$$\text{XOS to total sugar ratio} \ (\%) \ = \ \frac{XOS \ in \ the \ liquid \ hydrolysate \ after \ autohydrolysis \ (g)}{Total \ sugars \ in \ the \ liquid \ hydrolysate \ after \ autohydrolysis \ (g)} \times 100 \tag{2}$$

3. Results

3.1. Xylooligosaccharides Production Depending on the Changes of the Reaction Temperature

Hemicellulose fractions in SSB were released into the liquid hydrolysate in the form of both monomeric and oligomeric sugars after autohydrolysis (Table 2). Hemicellulosic sugars could not be fully hydrolyzed due to the severity of autohydrolysis, and a large amount of oligomeric sugars might be dissolved in the liquid hydrolysate together with monomeric sugars.

Table 2. Chemical composition of the liquid hydrolysate by reaction temperature variables.

Conditions		Sugars			Acetic Acid	Degradation Product	
Reaction Temperature (°C)	pH	Glucose	XGM [1]	Arabinose		HMF [2]	Furfural
160	4.1	3.2 ± 0.1	5.9 ± 0.9	2.0 ± 0.1	0.2 ± 0.0	ND [3]	ND
170	3.8	3.7 ± 0.1	12.2 ± 0.9	2.2 ± 0.1	0.4 ± 0.0	ND	ND
180	3.7	3.9 ± 0.1	16.9 ± 0.1	2.1 ± 0.1	0.8 ± 0.1	ND	0.2 ± 0.0
190	3.5	3.9 ± 0.0	17.9 ± 0.0	1.6 ± 0.1	1.5 ± 0.0	ND	0.8 ± 0.0
200	3.3	3.8 ± 0.1	13.9 ± 0.9	1.0 ± 0.1	2.4 ± 0.2	0.2 ± 0.0	2.3 ± 0.3

[1] Sum of xylose, galactose, and mannose; [2] Hydroxymethylfurfural; [3] Not detected.

The release trends of the three kinds of sugars were different depending on the changes in the reaction temperature. The XGM content in the liquid hydrolysate increased by increasing the reaction temperature to 18.4% based on the initial biomass, while the XGM content sharply decreased at temperatures above 200 °C. Based on the initial arabinan content as monomers, 84.6% of the arabinose could be released at 170 °C. The hemicellulosic sugars have good solubility in water after autohydrolysis, but the maximum release point was slightly different between XGM and arabinose. The glucose content showed a constant value of approximately 4% in the liquid hydrolysate. Compared to the initial glucan content in SSB, a small portion of glucose was released from SSB regardless of the reaction temperature changes. Thus, the autohydrolysis could selectively extract XGM in SSB, and the liquid hydrolysate generally had a large amount of XGM and a small amount of glucose and arabinose.

As the reaction temperature increased, the content of monomeric XGM dramatically increased due to the excess hydrolysis of hemicellulose (Figure 1). Meanwhile, XOS content was maximized at 180 °C (15.5%). Similar to the XOS content, the highest XOS to total sugar ratio was achieved at the same reaction temperature (180 °C). Therefore, an appropriate reaction temperature for improving the XOS ratio in the liquid hydrolysate could be found for SSB.

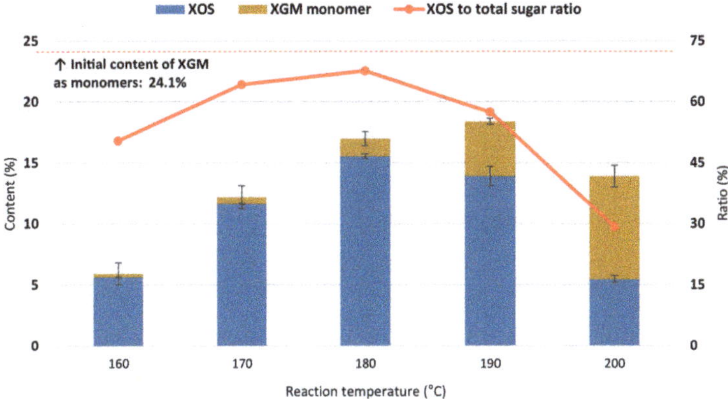

Figure 1. XGM content and XOS to total sugar ratio with the reaction temperature change.

3.2. Xylooligosaccharides Production Depending on the Changes of the Reaction Time

The autohydrolysis conditions for XOS production were investigated by changing the reaction time because it has been considered as one of the critical hydrothermal treatment parameters for the control of XOS DP. The reaction time was specified in 5-min intervals within 20 min to confirm the result in a shorter reaction time than that of the previous sections. The reaction temperature was set at 180 and 190 °C, which has a good performance for the high XOS ratio in the liquid hydrolysates.

The XGM content in the liquid hydrolysate was changed by an increase in the reaction time (Table 3). Both reaction temperature conditions presented an increasing trend of XGM content with longer reaction times, except for 20 min at 190 °C. The maximum XGM content was obtained at 180 °C for 20 min (16.9%) and 190 °C for 15 min (18.3%), respectively.

Table 3. Monomeric sugar composition of the liquid hydrolysate by reaction temperature and time variables.

Conditions		Sugars			Acetic Acid	Degradation Product	
Reaction Temperature (°C)	Reaction Time (min)	Glucose	XGM [1]	Arabinose		HMF [2]	Furfural
180	5	3.5 ± 0.2	11.3 ± 1.3	1.5 ± 1.1	0.4 ± 0.0	ND [3]	0.1 ± 0.0
	10	3.8 ± 0.1	13.6 ± 0.8	2.1 ± 0.1	0.6 ± 0.2	ND	0.1 ± 0.0
	15	3.9 ± 0.0	15.5 ± 0.9	2.1 ± 0.0	0.6 ± 0.1	ND	0.1 ± 0.0
	20	3.9 ± 0.1	16.9 ± 0.1	2.1 ± 0.1	0.8 ± 0.1	ND	0.2 ± 0.0
190	5	3.8 ± 0.0	15.6 ± 1.2	2.0 ± 0.0	0.7 ± 0.2	ND	0.1 ± 0.1
	10	3.8 ± 0.2	16.9 ± 0.7	1.9 ± 0.1	0.9 ± 0.1	ND	0.3 ± 0.1
	15	3.9 ± 0.0	18.3 ± 0.2	1.8 ± 0.1	1.2 ± 0.2	ND	0.5 ± 0.1
	20	3.9 ± 0.0	17.9 ± 0.0	1.6 ± 0.0	1.5 ± 0.0	ND	0.8 ± 0.0

[1] Sum of xylose, galactose, and mannose; [2] Hydroxymethylfurfural; [3] Not detected.

Although a good maximum XGM content was obtained at 190 °C than at 180 °C, the results of the XOS to total sugar ratio are quite different between the two conditions (Figure 2). In the cases of 180 °C, the XOS to total sugar ratio was constant even though the reaction time was changed (Figure 2a). Meanwhile, the ratio showed a sharp decrease as the reaction time was extended to 20 min at 190 °C (Figure 2b). Although 180 °C for 20 min showed the highest XOS content (15.4%), this condition could not guarantee the highest level of XOS to total sugar ratio (67.7%). Therefore, 190 °C for 5 min might be considered as the best point for XOS production from SSB due to the high content of XOS (14.7%) and high purity (XOS to total sugar ratio: 68.7%) than other autohydrolysis conditions.

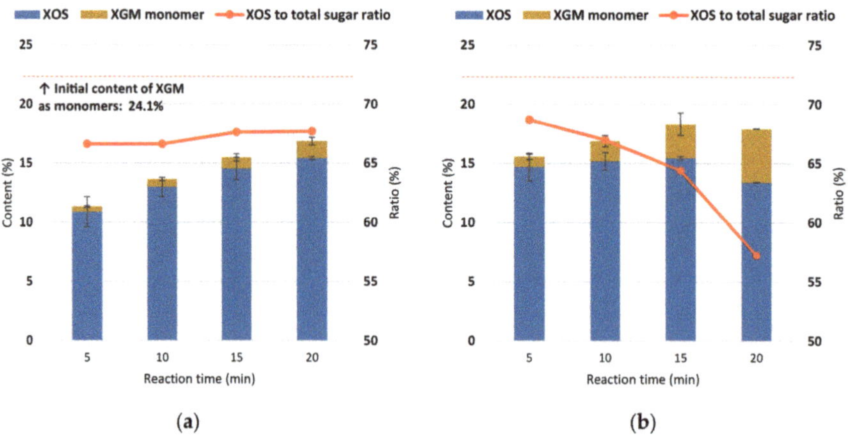

Figure 2. XGM content and XOS to total sugar ratio with reaction time change: (**a**) 180 °C; (**b**) 190 °C.

3.3. Glucose Conversion Depending on the Reaction Temparature Changes

The cellulose fraction in the solid residues after autohydrolysis was hydrolyzed to glucose by an enzyme cocktail (Figure 3). According to the sugar composition in the liquid hydrolysate, a small amount of glucose was released regardless of the changes in the reaction temperature (Table 1). However, the glucose yield after enzymatic hydrolysis showed an increasing trend with increasing reaction temperature. The glucose to total sugar ratio presented a similar trend with the glucose yield corresponding to the reaction temperature increase. Thus, the maximum glucose yield (36.6%) and glucose to total sugar ratio (94.2%) were obtained at 200 °C.

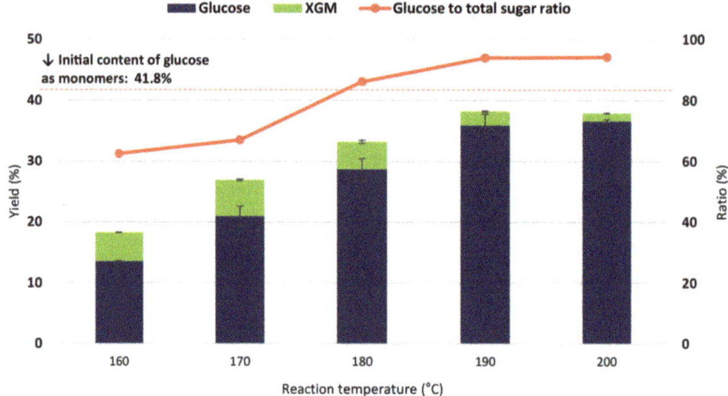

Figure 3. Glucose yield and glucose to total sugar ratio after enzymatic hydrolysis.

3.4. Glucose Conversion Depending on the Changes of the Reaction Time

The glucose yield was improved by extending the reaction time under both reaction temperature conditions (Figure 4). In addition, the glucose to total sugar ratio was maximized at 20 min, but the results were slightly different between the 180 °C (86.2%) and 190 °C (93.9%) conditions. However, it was revealed that a condition with a high glucose to total sugar ratio does not ensure a high XOS to total sugar ratio by comparing the results of the previous section.

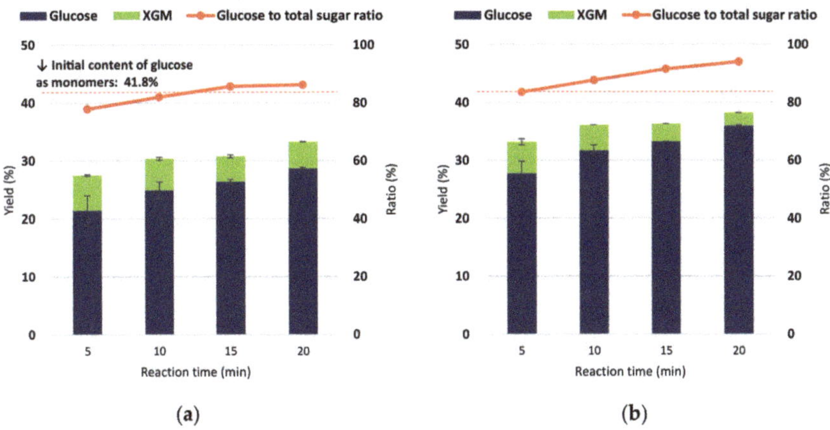

Figure 4. Glucose content and conversion ratio after enzymatic hydrolysis: (**a**) 180 °C; (**b**) 190 °C.

This trade-off relationship between XOS and glucose is obvious in the mass balance profile of this study (Figure 5). This figure suggests three autohydrolysis conditions: the first and second conditions represent the highest XOS to total sugar ratio at 180 and 190 °C, respectively. At 190 °C for 5 min, the XOS to total sugar ratio was higher (68.7%) than that at 180 °C for 20 min (67.6%). Meanwhile, the 180 °C condition has a superb result for the glucose to total sugars ratio (86.2%) than the 180 °C result (83.5%). Meanwhile, the 190 °C for 15 min condition presented a higher number of glucose to total sugar ratios (91.4%) compared to the two previous conditions, even though this condition had a moderate level of XOS to total sugar ratio (64.4%). Therefore, the autohydrolysis condition should be properly considered for optimizing the production of major products from SSB.

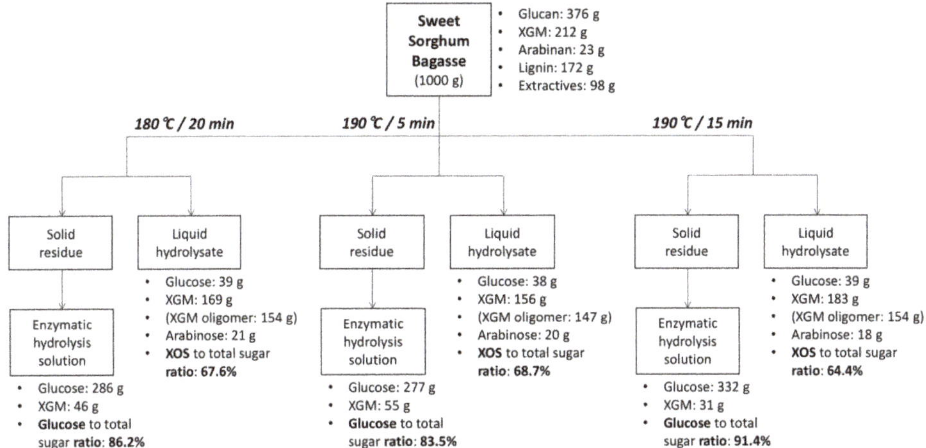

Figure 5. Mass balance of the XOS and glucose production from sweet sorghum bagasse.

4. Discussion

Numerous studies have attempted to improve the release performance of XOS from biomasses that directly connected the productivity of XOS [16–19]. However, control of impurities in the liquid fraction after treatment is considered a major factor affecting the market price of XOS as well as its yield [2,32]. Due to the fact that the purification process of XOS is costly, it can dramatically increase if a high amount of impurities are simultaneously mixed with XOS [28]. Another reason for the importance of controlling the impurities is related to the XOS activity [26]. In other words, the concept of impurity can be expanded to monomeric sugars, such as xylose [33]. XOS has several beneficial activities for the human body as a prebiotic, while xylose does not have any function like XOS [34]. Thus, preventing the excess generation of xylose is an essential factor for increasing the productivity of XOS.

The chemical composition of SSB showed good potential for XOS production (Table 1). It had a high amount of xylan, and a low amount of ash and other hemicellulosic sugars. In particular, a very tiny amount of galactan and mannan in SSB has been reported in previous studies [35,36]. Therefore, the XGM monomer in the liquid hydrolysate can be considered as xylose alone, and the XGM oligomer implied XOS in this study.

The acid-hydrolysis pathway of xylan is shown by the variation of the XGM content with an increase in the reaction temperature (Table 2). The xylan seemed to be fully degraded and dissolved in liquid hydrolysate at 190 °C. Then, the XGM content sharply dropped at 200 °C because it converted to HMF and furfural, which are degradation products. The pH level gradually decreased by providing more heat energy due to the cleavage of the acetyl group from the hemicellulose chain, which resulted in more acidic circumstances at higher reaction temperature conditions [37]. This means that xylan degradation might be accelerated by positive feedback from heat energy and acids during autohydrolysis. This phenomenon was revealed by the variation in the XGM monomer content, which increased more in the higher reaction temperature conditions (Figure 1). Thus, autohydrolysis proved that the xylan structure in SSB can be sufficiently decomposed without the addition of chemical catalysts.

As mentioned above, the quality of XOS might be controlled by reducing the impurities, including the monomeric sugars. In this term, the ratio between the XOS and the sum of glucose, arabinose, and XGM monomer in the liquid hydrolysate was calculated to evaluate the quality of XOS, and it was named XOS to total sugar ratio in this study. In other words, the XOS to total sugar ratio implies the proportion of XOS compared to the total sugars in the liquid hydrolysate. When the autohydrolysis conditions had a designated reaction time (20 min), the 170 °C condition showed a better XOS to total sugar ratio (64.2%) than that at 190 °C (57.5%) (Figure 1). The liquid hydrolysate at 170 °C might have fewer impurities than that at 190 °C, but the quantity of XOS and monomers at 190 °C (13.9% and 4.5%) was higher than that at 170 °C (11.6% and 0.5%). Considering the acid-hydrolysis pathway of xylan during autohydrolysis, the high amount of XGM monomer implies that the liquid hydrolysate at 190 °C could have a higher amount of low DP of XOS than that at 170 °C [38]. According to previous research, the low DP of XOS presented performed better as a prebiotic than high DP (above 4) of XOS [39].

180 °C, which had the highest XOS to total sugar ratio, and 190 °C conditions were selected for evaluating XOS production from SSB with several variations of the reaction time (Figure 3). The highest XOS content (15.4%) in this study was obtained at 180 °C for 20 min and 190 °C for 15 min. This XOS content can be converted to conversion ratio based on the initial amount of xylan in SSB (64.1%), which may be considered as a noteworthy value compared with the results of previous autohydrolysis studies: 60.9% from miscanthus [32], 61% from the brewery's spent grain [40], 51.7% from rice husks, and 51.8% from corn cob [41].

Glucan was presented as a major carbohydrate in SSB, even though it contained twice as much as the XGM in this study. Glucose has been recognized as a platform chemical from biomass and can be easily produced by enzymatic hydrolysis if the cell wall structure is appropriately degraded by the pretreatment [42]. The glucose to total sugar ratio, which was calculated in the same manner as the case of XOS, gradually increased with an increase in the reaction temperature and time (Figures 2 and 4). Meanwhile, less than 4% of glucose was released into the liquid hydrolysate even at the highest reaction temperature (Tables 2 and 3). This means that the amount of remaining glucan had not changed much by changing the autohydrolysis conditions, but the cell wall structure might be degraded toward a suitable form for cellulase activity at high reaction temperatures and long reaction times. The highest glucose to total sugar ratio (94.2%) was observed at 200 °C for 20 min, and an indication of glucan degradation could not be found under the autohydrolysis conditions in this study.

The ideal autohydrolysis condition for XOS production could be suggested by the variations in reaction temperature and time (Figure 5). A XOS to total sugar ratio of 68.7% was obtained at 190 °C for 5 min, and a short reaction time can be an advantage for large-scale plants [43]. However, the result of this condition is concerned that the XOS might consist of a significant amount of high DP of XOS. According to this assumption, the liquid hydrolysate under this condition may require further processing to reduce the DP by xylanase. Additionally, this condition showed a relatively low glucose to total sugar ratio (83.5%), another main product of autohydrolysis, than other conditions. In this term, 180 °C for 20 min can be an alternative to 190 °C for 5 min because the XOS to total sugar ratio in this condition (67.6%) is slightly lower than that at 190 °C for 5 min. Meanwhile, the higher XGM monomer content (1.4%) in this condition means that the average DP might be lower than that at 190 °C for 5 min. If xylanase treatment followed by autohydrolysis induces the formation of XOS with a narrow DP from 2 to 4, the XOS DP variation in the liquid hydrolysate should be considered as an important factor in determining the amount of xylanase consumed. In this regard, 180 °C for 20 min might be more favorable for producing high-value XOS than at 190 °C for 5 min. On the other hand, 190 °C for 15 min can be chosen to obtain a high purity of glucose, 91.4% of glucose to total sugar ratio, even though this condition showed a moderate level of XOS to total sugar ratio (64.4%). This trade-off relationship between XOS and glucose might be controlled by inducing an additional process between the autohydrolysis and enzymatic hydrolysis steps. For instance, mechanical refining can be employed for solid residues, and it improves glucose conversion by collapsing the cell wall structure, even though the solid residue is produced under low reaction temperature conditions [44].

5. Conclusions

The xylan-rich edible biomass, SSB, was utilized for XOS and glucose production using the autohydrolysis process. The purity of XOS was improved by controlling the autohydrolysis conditions, and XOS to total sugar ratio was maximized by up to 68.7%, which is expected to reduce production costs for the purification process. However, both XOS purity and content should be considered to determine the optimum conditions for XOS production. On the other hand, the remaining solid residue after autohydrolysis was successfully converted to glucose by enzymatic hydrolysis, and the highest glucose to total sugar ratio was 94.2%. Although there was a discrepancy between the ideal conditions for XOS and glucose production, the results of this study can be useful in selecting the suitable autohydrolysis condition for improving the production of the desired material from SSB.

Author Contributions: Conceptualization, H.K.; methodology, C.-D.J. and H.K.; validation, H.K.; formal analysis, C.-D.J.; investigation, C.-D.J. and H.K.; resources J.-H.Y. and H.K.; data curation, S.-K.J. and H.K.; writing—original draft preparation, S.-K.J.; writing—review and editing, S.-K.J. and H.K.; visualization, S.-K.J.; supervision, H.K.; project administration, H.K.; funding acquisition, H.K. All authors have read and agreed to the published version of the manuscript.

Funding: This work was supported by the Korea Research Institute of Chemical Technology (KRICT, Korea) project (SS2042-10) and R&D Program (20008416) of the Ministry of Trade, Industry & Energy (MOTIE/KEIT, Korea).

Conflicts of Interest: The authors declare no conflict of interest.

References

1. Amorim, C.; Silvério, S.C.; Prather, K.L.; Rodrigues, L.R. From lignocellulosic residues to market: Production and commercial potential of xylooligosaccharides. *Biotechnol. Adv.* **2019**, *37*, 107397. [CrossRef] [PubMed]
2. Aachary, A.A.; Prapulla, S.G. Xylooligosaccharides (XOS) as an emerging prebiotic: Microbial synthesis, utilization, structural characterization, bioactive properties, and applications. *Compr. Rev. Food Sci. Food Saf.* **2011**, *10*, 2–16. [CrossRef]

3. Bhatia, R.; Winters, A.; Bryant, D.N.; Bosch, M.; Clifton-Brown, J.; Leak, D.; Gallagher, J. Pilot-scale production of xylo-oligosaccharides and fermentable sugars from Miscanthus using steam explosion pretreatment. *Bioresour. Technol.* **2020**, *296*, 122285. [CrossRef] [PubMed]

4. Swennen, K.; Courtin, C.M.; Delcour, J.A. Non-digestible oligosaccharides with prebiotic properties. *Crit. Rev. Food Sci. Nutr.* **2006**, *46*, 459–471. [CrossRef] [PubMed]

5. Singh, R.; Banerjee, J.; Sasmal, S.; Muir, J.; Arora, A. High xylan recovery using two stage alkali pre-treatment process from high lignin biomass and its valorisation to xylooligosaccharides of low degree of polymerisation. *Bioresour. Technol.* **2018**, *256*, 110–117. [CrossRef]

6. Xiao, X.; Bian, J.; Peng, X.-P.; Xu, H.; Xiao, B.; Sun, R.-C. Autohydrolysis of bamboo (Dendrocalamus giganteus Munro) culm for the production of xylo-oligosaccharides. *Bioresour. Technol.* **2013**, *138*, 63–70. [CrossRef]

7. Xylooligosaccharides (XOS) Market Size 2020 Analysis by CAGR of 4.1%, I.S., Business Strategies, Emerging Demands, Growth Rate, Recent Trends, Opportunity, and Forecast to 2026. Available online: https://www.marketwatch.com/press-release/xylooligosaccharides-xos-market-size-2020-analysis-by-cagr-of-41-industry-share-business-strategies-emerging-demands-growth-rate-recent-trends-opportunity-and-forecast-to-2026-2020-07-12 (accessed on 20 September 2020).

8. Bian, J.; Peng, F.; Peng, X.-P.; Peng, P.; Xu, F.; Sun, R.-C. Structural features and antioxidant activity of xylooligosaccharides enzymatically produced from sugarcane bagasse. *Bioresour. Technol.* **2013**, *127*, 236–241. [CrossRef]

9. Zhang, W.; You, Y.; Lei, F.; Li, P.; Jiang, J. Acetyl-assisted autohydrolysis of sugarcane bagasse for the production of xylo-oligosaccharides without additional chemicals. *Bioresour. Technol.* **2018**, *265*, 387–393. [CrossRef]

10. Chen, H.; Liu, J.; Chang, X.; Chen, D.; Xue, Y.; Liu, P.; Lin, H.; Han, S. A review on the pretreatment of lignocellulose for high-value chemicals. *Fuel Process. Technol.* **2017**, *160*, 196–206. [CrossRef]

11. Qing, Q.; Li, H.; Kumar, R.; Wyman, C.E. Xylooligosaccharides production, quantification, and characterization in context of lignocellulosic biomass pretreatment. *Aqueous Pretreat. Plant Biomass Biol. Chem. Convers. Fuels Chem.* **2013**, 391–415.

12. Poletto, P.; Pereira, G.N.; Monteiro, C.R.; Pereira, M.A.F.; Bordignon, S.E.; de Oliveira, D. Xylooligosaccharides: Transforming the lignocellulosic biomasses into valuable 5-carbon sugar prebiotics. *Process Biochem.* **2020**, *91*, 352–363. [CrossRef]

13. Garrote, G.; Domínguez, H.; Parajó, J.C. Mild autohydrolysis: An environmentally friendly technology for xylooligosaccharide production from wood. *J. Chem. Technol. Biotechnol.* **1999**, *74*, 1101–1109. [CrossRef]

14. Lachos-Perez, D.; Martinez-Jimenez, F.; Rezende, C.; Tompsett, G.; Timko, M.; Forster-Carneiro, T. Subcritical water hydrolysis of sugarcane bagasse: An approach on solid residues characterization. *J Supercrit. Fluids* **2016**, *108*, 69–78. [CrossRef]

15. Kumar, L.; Tooyserkani, Z.; Sokhansanj, S.; Saddler, J.N. Does densification influence the steam pretreatment and enzymatic hydrolysis of softwoods to sugars? *Bioresour. Technol.* **2012**, *121*, 190–198. [CrossRef] [PubMed]

16. Carvalho, A.F.A.; Marcondes, W.F.; de Oliva Neto, P.; Pastore, G.M.; Saddler, J.N.; Arantes, V. The potential of tailoring the conditions of steam explosion to produce xylo-oligosaccharides from sugarcane bagasse. *Bioresour. Technol.* **2018**, *250*, 221–229. [CrossRef]

17. Oliveira, F.M.; Pinheiro, I.O.; Souto-Maior, A.M.; Martin, C.; Gonçalves, A.R.; Rocha, G.J. Industrial-scale steam explosion pretreatment of sugarcane straw for enzymatic hydrolysis of cellulose for production of second generation ethanol and value-added products. *Bioresour. Technol.* **2013**, *130*, 168–173. [CrossRef]

18. Liu, Z.-H.; Qin, L.; Pang, F.; Jin, M.-J.; Li, B.-Z.; Kang, Y.; Dale, B.E.; Yuan, Y.-J. Effects of biomass particle size on steam explosion pretreatment performance for improving the enzyme digestibility of corn stover. *Ind. Crops Prod.* **2013**, *44*, 176–184. [CrossRef]

19. Álvarez, C.; González, A.; Negro, M.J.; Ballesteros, I.; Oliva, J.M.; Sáez, F. Optimized use of hemicellulose within a biorefinery for processing high value-added xylooligosaccharides. *Ind. Crops Prod.* **2017**, *99*, 41–48. [CrossRef]

20. Menandro, L.M.S.; Cantarella, H.; Franco, H.C.J.; Kölln, O.T.; Pimenta, M.T.B.; Sanches, G.M.; Rabelo, S.C.; Carvalho, J.L.N. Comprehensive assessment of sugarcane straw: Implications for biomass and bioenergy production. *Biofuel Bioprod. Bior.* **2017**, *11*, 488–504. [CrossRef]

21. Silva, T.A.L.; Zamora, H.D.Z.; Varão, L.H.R.; Prado, N.S.; Baffi, M.A.; Pasquini, D. Effect of steam explosion pretreatment catalysed by organic acid and alkali on chemical and structural properties and enzymatic hydrolysis of sugarcane bagasse. *Waste Biomass Valorization* **2018**, *9*, 2191–2201. [CrossRef]

22. Reddy, B.V.; Kumar, A.A.; Ramesh, S. *Sweet Sorghum: A Water Saving Bio-Energy Crop*; ICRISAT: Patancheru, India, 2007.

23. Erickson, J.E.; Woodard, K.R.; Sollenberger, L.E. Optimizing sweet sorghum production for biofuel in the southeastern USA through nitrogen fertilization and top removal. *Bioenergy Res.* **2012**, *5*, 86–94. [CrossRef]

24. Koo, B.; Park, J.; Gonzalez, R.; Jameel, H.; Park, S. Two-stage autohydrolysis and mechanical treatment to maximize sugar recovery from sweet sorghum bagasse. *Bioresour. Technol.* **2019**, *276*, 140–145. [CrossRef] [PubMed]

25. Hughes, S.; Shewry, P.; Li, L.; Gibson, G.; Sanz, M.; Rastall, R. In vitro fermentation by human fecal microflora of wheat arabinoxylans. *J. Agric. Food. Chem.* **2007**, *55*, 4589–4595. [CrossRef] [PubMed]

26. Ho, A.L.; Kosik, O.; Lovegrove, A.; Charalampopoulos, D.; Rastall, R.A. In vitro fermentability of xylo-oligosaccharide and xylo-polysaccharide fractions with different molecular weights by human faecal bacteria. *Carbohydr. Polym.* **2018**, *179*, 50–58. [CrossRef] [PubMed]

27. Ruiz, E.; Gullón, B.; Moura, P.; Carvalheiro, F.; Eibes, G.; Cara, C.; Castro, E. Bifidobacterial growth stimulation by oligosaccharides generated from olive tree pruning biomass. *Carbohydr. Polym.* **2017**, *169*, 149–156. [CrossRef]

28. Otieno, D.O.; Ahring, B.K. A thermochemical pretreatment process to produce xylooligosaccharides (XOS), arabinooligosaccharides (AOS) and mannooligosaccharides (MOS) from lignocellulosic biomasses. *Bioresour. Technol.* **2012**, *112*, 285–292. [CrossRef]

29. Sluiter, A.; Hames, B.; Ruiz, R.; Scarlata, C.; Sluiter, J.; Templeton, D. Determination of Sugars, Byproducts, and Degradation Products in Liquid Fraction Process Samples. 2006. Available online: https://www.nrel.gov/docs/gen/fy08/42623.pdf (accessed on 27 October 2020).

30. Sluiter, A.; Hames, B.; Ruiz, R.; Scarlata, C.; Sluiter, J.; Templeton, D.; Crocker, D. Determination of structural carbohydrates and lignin in biomass. *Lab. Anal. Proced.* **2008**, *1617*, 1–16.

31. Salo-väänänen, P.P.; Koivistoinen, P.E. Determination of protein in foods: Comparison of net protein and crude protein (N× 6.25) values. *Food Chem.* **1996**, *57*, 27–31. [CrossRef]

32. Chen, M.-H.; Bowman, M.J.; Dien, B.S.; Rausch, K.D.; Tumbleson, M.; Singh, V. Autohydrolysis of Miscanthus x giganteus for the production of xylooligosaccharides (XOS): Kinetics, characterization and recovery. *Bioresour. Technol.* **2014**, *155*, 359–365. [CrossRef]

33. Gullón, P.; Moura, P.; Esteves, M.a.P.; Girio, F.M.; Domínguez, H.; Parajó, J.C. Assessment on the fermentability of xylooligosaccharides from rice husks by probiotic bacteria. *J. Agric. Food. Chem.* **2008**, *56*, 7482–7487. [CrossRef]

34. Moura, P.; Barata, R.; Carvalheiro, F.; Gírio, F.; Loureiro-Dias, M.C.; Esteves, M.P. In vitro fermentation of xylo-oligosaccharides from corn cobs autohydrolysis by Bifidobacterium and Lactobacillus strains. *LWT* **2007**, *40*, 963–972. [CrossRef]

35. Sun, S.; Wen, J.; Sun, S.; Sun, R.-C. Systematic evaluation of the degraded products evolved from the hydrothermal pretreatment of sweet sorghum stems. *Biotechnol. Biofuels* **2015**, *8*, 37. [CrossRef] [PubMed]

36. Pham, H.T.T.; Nghiem, N.P.; Kim, T.H. Near theoretical saccharification of sweet sorghum bagasse using simulated green liquor pretreatment and enzymatic hydrolysis. *Energy* **2018**, *157*, 894–903. [CrossRef]

37. Wei, H.; Chen, X.; Shekiro, J.; Kuhn, E.; Wang, W.; Ji, Y.; Kozliak, E.; Himmel, M.E.; Tucker, M.P. Kinetic modelling and experimental studies for the effects of Fe2+ ions on xylan hydrolysis with dilute-acid pretreatment and subsequent enzymatic hydrolysis. *Catalysts* **2018**, *8*, 39. [CrossRef]

38. Surek, E.; Buyukkileci, A.O. Production of xylooligosaccharides by autohydrolysis of hazelnut (*Corylus avellana* L.) shell. *Carbohydr. Polym.* **2017**, *174*, 565–571. [CrossRef] [PubMed]

39. Childs, C.E.; Röytiö, H.; Alhoniemi, E.; Fekete, A.A.; Forssten, S.D.; Hudjec, N.; Lim, Y.N.; Steger, C.J.; Yaqoob, P.; Tuohy, K.M. Xylo-oligosaccharides alone or in synbiotic combination with *Bifidobacterium animalis* subsp. lactis induce bifidogenesis and modulate markers of immune function in healthy adults: A double-blind, placebo-controlled, randomised, factorial cross-over study. *Br. J. Nutr.* **2014**, *111*, 1945–1956. [CrossRef]

40. Carvalheiro, F.; Esteves, M.; Parajó, J.; Pereira, H.; Gırio, F. Production of oligosaccharides by autohydrolysis of brewery's spent grain. *Bioresour. Technol.* **2004**, *91*, 93–100. [CrossRef]

41. Garrote, G.; Falqué, E.; Domínguez, H.; Parajó, J.C. Autohydrolysis of agricultural residues: Study of reaction byproducts. *Bioresour. Technol.* **2007**, *98*, 1951–1957. [CrossRef]

42. Zakaria, M.R.; Hirata, S.; Hassan, M.A. Hydrothermal pretreatment enhanced enzymatic hydrolysis and glucose production from oil palm biomass. *Bioresour. Technol.* **2015**, *176*, 142–148. [CrossRef]

43. Jeon, H.; Kang, K.-E.; Jeong, J.-S.; Gong, G.; Choi, J.-W.; Abimanyu, H.; Ahn, B.S.; Suh, D.-J.; Choi, G.-W. Production of anhydrous ethanol using oil palm empty fruit bunch in a pilot plant. *Biomass Bioenergy* **2014**, *67*, 99–107. [CrossRef]

44. Park, J.; Jones, B.; Koo, B.; Chen, X.; Tucker, M.; Yu, J.-H.; Pschorn, T.; Venditti, R.; Park, S. Use of mechanical refining to improve the production of low-cost sugars from lignocellulosic biomass. *Bioresour. Technol.* **2016**, *199*, 59–67. [CrossRef] [PubMed]

Publisher's Note: MDPI stays neutral with regard to jurisdictional claims in published maps and institutional affiliations.

applied
sciences

Article

Catalytic Conversion of α-Pinene to High-Density Fuel Candidates Over Stannic Chloride Molten Salt Hydrates

Seong-Min Cho [1], June-Ho Choi [1], Jong-Hwa Kim [1], Bonwook Koo [2] and In-Gyu Choi [3,4,*]

[1] Department of Forest Sciences, Seoul National University, Seoul 08826, Korea; csmin93@snu.ac.kr (S.-M.C.); jhchoi1990@snu.ac.kr (J.-H.C.); wmfty@snu.ac.kr (J.-H.K.)

[2] Green and Sustainable Materials R & D Department, Korea Institute of Industrial Technology, Cheonan 31056, Korea; bkoo@kitech.re.kr

[3] Department of Agriculture, Forestry and Bioresources, Seoul National University, Seoul 08826, Korea

[4] Research Institute of Agriculture and Life Sciences, Seoul National University, Seoul 08826, Korea

* Correspondence: cingyu@snu.ac.kr; Tel.: +82-2-880-4785

Received: 18 September 2020; Accepted: 22 October 2020; Published: 26 October 2020

Abstract: The synthesis of dimeric products from monoterpene hydrocarbons has been studied for the development of renewable high-density fuel. In this regard, the conversion of α-pinene in turpentine over stannic chloride molten salt hydrates ($SnCl_4 \cdot 5H_2O$) as a catalyst was investigated, and the reaction products were analyzed with gas chromatography/flame ionization detector/mass spectrometer (GC/FID/MS). Overall, the content of α-pinene in a reaction mixture decreased precipitously with an increasing reaction temperature. Almost 100% of the conversion was shown after 1 h of reaction above 90 °C. From α-pinene, dimeric products (hydrocarbons and alcohols/ethers) were mostly formed and their yield showed a steady increase of up to 61 wt% based on the reaction mixture along with the reaction temperature. This conversion was thought to be promoted by Brønsted acid activity of the catalyst, which resulted from a Lewis acid-base interaction between the stannic ($Sn^{(IV)}$) center and the coordinated water ligands. As for the unexpected heteroatom-containing products, oxygen and chlorine atoms were originated from the coordinated water and chloride ligands of the catalyst. Based on the results, we constructed not only a plausible catalytic cycle of $SnCl_4 \cdot 5H_2O$ but also the mechanism of catalyst decomposition.

Keywords: renewable fuel; high-density fuel; α-pinene dimerization; turpentine; stannic chloride molten salt hydrates

---

1. Introduction

Aspirations for alternative biofuel have emerged vigorously worldwide, amid increasing concern about fossil fuel sustainability and environmental pollution. In this respect, cellulosic biomass-derived chemicals such as furfural, 5-hydroxymethylfurfural, 2-methylfuran, levulinic esters, and angelica lactone have attracted considerable attention as renewable resources for the sustainable production of biofuels [1–3]. Another promising renewable chemical in order to produce synthetic fuel is turpentine. Turpentine can be obtained by either the distillation of oleoresins or the kraft pulping process as a side-product and it is mainly composed of α-pinene and lesser amounts of isomers (β-pinene, camphene, and limonene), which all belong to monoterpene hydrocarbons [4]. Because fuel properties depend on what consists of the fuel, the fact that hydrocarbons make up the majority of turpentine is an advantage of itself as the resource of biofuels [5]. It has been reported that the blends of monoterpenes and petroleum-based fuel could be utilized as drop-in fuels [6,7].

When it comes to the application of fuel to an aerospace field and ramjet engines, however, drop-in fuels are required for much higher energy density than neat turpentine. To synthesize high-energy density fuel from turpentine (α-pinene), dimerization over acid catalysts has been devised to increase density, thereby increasing energy density [5]. With the advantage of facile separation, various heterogeneous Brønsted acid catalysts have been studied to carry out the dimerization [8,9]. Because the four-membered ring of α-pinene is labile to acidic conditions, the ring is easily opened or extended to five-membered ring. This furnishes isomers including both monocyclic monoterpene hydrocarbons such as limonene and bicyclic monoterpene hydrocarbons such as camphene. Thus, the acid-catalyzed dimerization of α-pinene always accompanies the isomerization, which is followed by inevitable cross-dimerization and makes the reaction mixture consist of not only the isomers of α-pinene but also the various dimers [8]. It has been found that a dimer fraction has too high viscosity to apply it to a field where the fuel is exposed to extremely low temperatures [10]. To come up with this problem, blending the dimer faction with monoterpene hydrocarbons or the partial dimerization of α-pinene has been devised [5].

Inorganic molten salts hydrates, such as $ZnCl_2 \cdot 3H_2O$, are the hydrated form of inorganic salts that easily melted like butter at an elevated temperature. A chemical reaction system exploiting the inorganic molten salts hydrates as a reaction medium has been applied to several processes, particularly biopolymer conversion. It has been reported that an acidic LiBr salt hydrates can hydrolyze both the β-1,4 glycosidic bond of cellulose [11] and the aryl ether bond of lignin [12] and co-produce boromomethylfurfural and furfural directly from lignocellulosic biomass within the biphasic system [13]. In addition to LiBr salt hydrates, $ZnCl_2$ and $SnCl_4$ salts hydrates have been used for the synthesis of isosorbide from cellulose [14] and 5-hydroxymethylfurfural from fructose, inulin, glucose, and chitosan [15–18].

Herein, we focused on the acidic nature of stannic chloride molten salt hydrates ($SnCl_4 \cdot 5H_2O$). $[SnCl_4L_2]$ complexes in anhydrous conditions have been reported as Lewis acid-assisted Brønsted acid catalysts when the L_2 are 2′-methoxy-[1,1′-binaphthalen]-2-ol (BINOL-Me), 2′-isopropoxy-[1,1′-binaphthalen]-2-ol (BINOL-*i*Pr), 2-methoxyphenol (guaiacol), or 2,6-dimethoxyphenol (syringol), because they can coordinate to the stannic ($Sn^{(IV)}$) center via the vicinal hydroxy and alkoxy groups in a bidentate manner [19–22]. Therefore, we have expected that the coordinated water ligands of $SnCl_4 \cdot 5H_2O$ can also act as Lewis acid-assisted Brønsted acids for the catalytic conversion of α-pinene. It has been remarkable that $SnCl_4 \cdot 5H_2O$ can furnish the dimerized products (up to 61 wt% based on the reaction mixture) from turpentine in a solvent-less condition. However, the decomposition of the catalyst results in oxygen-containing or chloride-containing products. To overcome this limitation and ponder the improvement of the reaction system, we have discussed the mechanism of the catalytic conversion of α-pinene over $SnCl_4 \cdot 5H_2O$.

2. Materials and Methods

2.1. Materials

All purchased chemicals were used without further purification. Turpentine (α-pinene ≥ 92% by GC) was purchased from Junsei Chemical Co., Ltd. (Tokyo, Japan). Stannic chloride molten salt hydrates ($SnCl_4 \cdot 5H_2O$, 98%) and n-hexane (95%) were obtained via Samchun Pure Chemicals Co., Ltd. (Seoul, Korea).

2.2. General Procedure for Catalytic Conversion of α-pinene by Stannic Chloride Molten Salt Hydrates

We set reaction time as 1 h and considered the effect of the reaction temperature (60, 70, 80, 90, 100, and 110 °C) or set reaction temperature as 70 °C and considered the effect of reaction time (1, 3, and 5 h) because, upon the reaction temperature crossed, the melting point of stannic chloride molten salt hydrates ($SnCl_4 \cdot 5H_2O$), α-pinene, quickly disappeared from the reaction mixture. In a typical experiment, 10 g of turpentine and 1 g of $SnCl_4 \cdot 5H_2O$ was added to 50-mL double-neck round flask equipped with a magnetic Teflon coated stirrer, a thermometer, and a reflux condenser. After being purged with nitrogen gas, the reactor was then loaded on a preheated aluminum heating block and stirred smoothly for 1 h. The temperature of a reaction mixture was reached at the desired temperature (±3 °C) in less than 5 min. Upon completion of the reaction, the reactor was then removed from the heating block and immediately cooled to room temperature with a water bath. After cooling, the crude reaction mixture was diluted in 100 mL of n-hexane and filtered over a celite pad to remove metal compounds. The solvent was removed under reduced pressure. The purified product was analyzed by Agilent Technologies 7890B gas chromatograph equipped with a DB-5ms column (30 m × 250 μm, 0.25 μm thickness), a 5977A mass spectrometer detector, and a flame ionization detector (Agilent-Technologies, Seoul, Korea). The yield of products was calculated based on the below equation on the basis of the internal standard method (tridecane).

$$\text{Yield of product } i \text{ (\%)} \; = \; \frac{\text{Weight of product } i \text{ after reaction}}{\text{Initial weight of } \alpha - \text{pinene}} \times 100 \tag{1}$$

3. Results and Discussion

3.1. Catalytic Conversion of α-Pinene over Stannic Chloride Molten Salt Hydrates

As previously mentioned in the introduction, what we have expected was that the Lewis acid-assisted Brønsted acidic nature of stannic chloride molten salt hydrates ($SnCl_4 \cdot 5H_2O$) would work as a catalyst. In accordance with this expectation, the catalytic amount of $SnCl_4 \cdot 5H_2O$ was active for the conversion of α-pinene in a solvent-less condition. As shown in Figure S1, the reaction products could be divided into four main groups (monomeric hydrocarbon, heteroatom-containing monomeric, dimeric, and trimeric product regions) on the basis of retention time in gas chromatography/flame ionization detector (GC/FID) chromatogram and molecular ion peaks in mass spectra. Figure S2A,B show the representative mass spectra of dimeric and trimeric hydrocarbons including molecular ion peaks at m/z 272 $[C_{20}H_{32}]^+$ and m/z 408 $[C_{30}H_{48}]^+$, respectively.

Figure 1A shows that α-pinene, which is the main constituent of turpentine, was consumed precipitously with increasing reaction temperature and the content thereof eventually stands at 0% in the reaction mixture heated above 90 °C. For the cases of dimeric and trimeric products, the contents increased approximately four-fold and eight-fold when the reaction temperature increased from 60 to 110 °C, standing at 60.9% and 15.4%, respectively, at 110 °C. There was a large discrepancy in the conversion of α-pinene and yields of dimeric and trimeric products, which denotes the presence of other monomeric products formed from α-pinene. Remarkably, our results differ from those of Nie et al. who suggested that, when phosphotungstic acid supported on MCM-41 was applied, the isomerization of α-pinene is the main reaction pathway at 100 °C and the dimeric products are formed mainly from the isomers at 160 °C [23]. As shown in Figure 1B, even at 70 °C of reaction temperature, 100% conversion of α-pinene can be achieved by increasing the reaction time. Given that the dimeric products were produced without α-pinene after 3 h of reaction, it can be thought that the monomeric products from α-pinene also participated in the dimerization [8].

Unfortunately, several mass spectra involving an unexpected molecular ion peak at $m/z = 290$ (Figure S2C) were extracted from the dimeric product region in Figure 1B. Given the difference between the m/z of molecular ions, the molecular ion having $m/z = 290$ could be described as $[C_{20}H_{34}O]^+$, which means that oxygen-containing dimeric products, possibly dimeric alcohols or ethers, were also produced during the reaction. In the dimerization of monoterpenes over acidic catalysts, dimeric products having $m/z = 290$ as a molecular ion have often been reported. Nie et al. suggested a dimeric alcohol product in the dimerization of crude turpentine [23]. Moreover, heteropoly acid-catalyzed conversion of camphene furnished diisobornyl ether in a moderate yield and good selectivity [24].

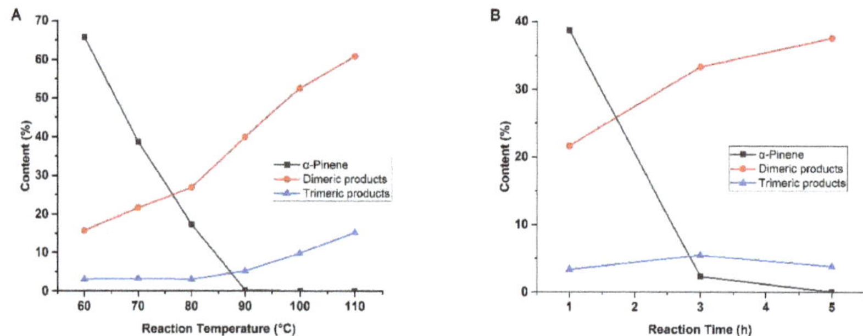

Figure 1. Effect of reaction temperature and reaction time on the conversion of α-pinene and yields of dimeric and trimeric products. The reaction was carried out (**A**) for 1 h and (**B**) at 70 °C.

Figure 2A,B show the main hydrocarbon product distribution in the monomeric hydrocarbon region in Figure S1B as a function of the reaction temperature. These monomeric hydrocarbons could be divided into two groups based on the tendency in the content thereof, according to the reaction temperature. Limonene, terpinolene, α-terpinene, γ-terpinene, and camphene except for tricyclene presented a sudden decrease in their yield above 90 °C, meaning their participation in dimerization. On the other hand, the contents of menthenes and *p*-cymene increased with increasing reaction temperature until 100 °C. These have been reported to form by the disproportionation of *p*-menthadienes [25]. In our results, the similar contents of menthenes and *p*-cymene also support the disproportionation mechanism. Meanwhile, as the same as the case of dimeric products, heteroatom-containing monomeric products were also furnished (Figure 2C). Non-negligible amounts of borneol, bornyl chloride, and fenchyl chloride were only produced as bicyclic compounds during the reaction. Figure 2D–F show the distribution of monomeric products with increasing reaction time at 70 °C of the reaction temperature. The considerable amounts of limonene (18.9%) and camphene (10.5%) were produced when α-pinene still remained in the reaction system whereas their yields decreased after the conversion of α-pinene was complete, which denotes their participation in dimerization.

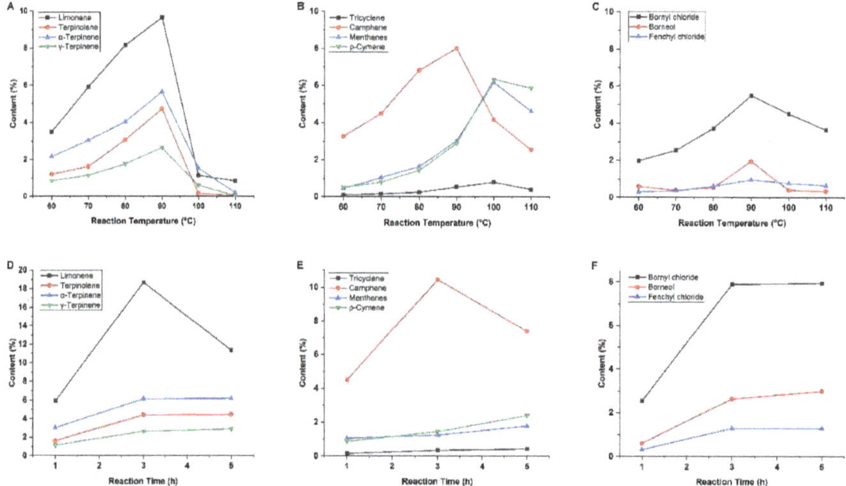

Figure 2. Effect of reaction temperature and reaction time on the yields of (**A,D**) *p*-menthadiene isomers, (**B,E**) other monomeric hydrocarbons, and (**C,F**) heteroatom-containing monomeric products. The reaction was carried out (**A–C**) for 1 h and (**D–F**) at 70 °C.

3.2. Rationalization of the Formation of Monoterpene Isomers and Dimeric Hydrocarbons

There have been a few papers on the $SnCl_4 \cdot 5H_2O$-catalyzed organic reactions. Based on the crystal structure of $SnCl_4 \cdot 5H_2O$ that consists of $[SnCl_4(H_2O)_2]$ linked by hydrogen bonding with additional water molecules [26], the mechanism of those reactions has been explained in the manner of both Lewis and Brønsted acid-base interaction. When Lewis basic heteroatom-containing reactants are used, their coordination to stannic ($Sn^{(IV)}$) center is considered to be an important step of the reaction. Such examples involve β-amino alcohol synthesis by epoxide ring-opening [27] and β-amino carbonyl compound synthesis by the Mannich-type reaction [28]. On the other hand, in the absence of heteroatom in reactants, the Brønsted acidity of the coordinated water ligands is considered to be important. For example, the hydration of alkyne has been explained by the initial protonation of reactants by water, which is activated by the coordination to the stannic center [29]. Such Lewis acid-assisted Brønsted acids were also developed for enantioselective polyene cyclization [19,20] and enantioselective protonation of silyl enol [21,22]. Given α-pinene has no heteroatom that can directly interact with the stannic center in a Lewis acid-base manner, the catalytic conversion of α-pinene by $SnCl_4 \cdot 5H_2O$ should follow the second explanation above.

The possible mechanisms for the conversion of α-pinene to isomerized products (tricyclene, camphene, limonene, terpinolene, α-terpinene, and γ-terpinene), disproportionated products (menthenes and *p*-cymene), and dimerized products by various acid catalysts have been proposed in the previous papers [8,23]. In this work, we focused on a plausible mechanism involving interaction between α-pinene and the catalyst, which is shown in Scheme 1.

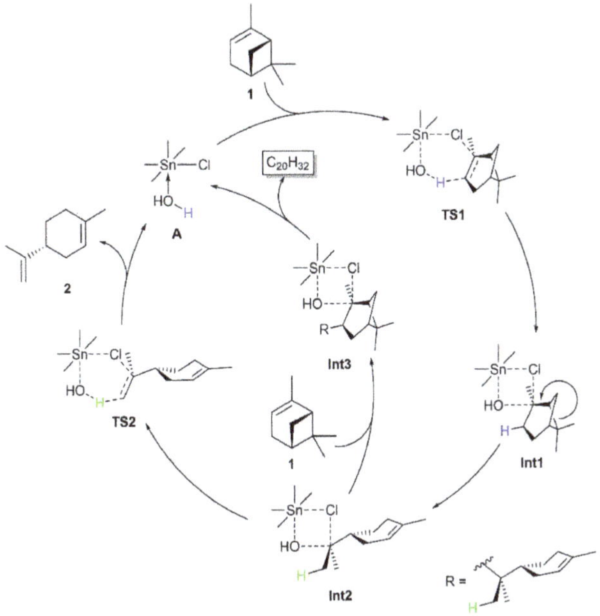

Scheme 1. A plausible catalytic cycle of $SnCl_4 \cdot 5H_2O$ (**A**) proposed for the synthesis of limonene (**2**) and dimeric hydrocarbons ($C_{20}H_{32}$) from α-pinene (**1**).

First, the protonation of α-pinene (1) is thought to be promoted by the Lewis acid-assisted Brønsted acid activity of the coordinated water ligands in the $Sn^{(IV)}$ complex (A) via TS1 (Scheme 1). Since there is no polar solvent that can solvate the generated ions, an intimate ion-pair between carbocation and $Sn^{(IV)}$ ate complex should be a reaction intermediate (Int1) [19,30]. To form limonene (2), the main isomerized product from compound 1, the pinanyl moiety in the intimate ion-pair Int1 should suffer from a ring-opening rearrangement, which generates a carbocation/Sn$^{(IV)}$ ate complex ion-pair (Int2). In this intimate ion-pair Int2, the abstraction of a proton from the carbocation moiety by the hydroxo ligand via TS2 regenerates stannic complex A and furnishes compound 2 as the product of a catalytic cycle. If the insertion of another compound 1 into the intimate ion-pair Int2 is prior to the proton abstraction step, a carbocation/Sn$^{(IV)}$ ate complex ion-pair (Int3) appears for dimeric hydrocarbons ($C_{20}H_{32}$) as the product of the catalytic cycle. Along with the case of compound 2, we can construct distinct routes to complete the catalytic cycle involving the possible combination of carbocation/Sn$^{(IV)}$ ate complex ion-pairs, which are able to explain the various isomers and dimeric hydrocarbons from compound 1.

3.3. Rationalization of the Formation of Heteroatom-Containing Dimeric Products

As previously mentioned, from α-pinene, stannic chloride molten salt hydrates ($SnCl_4 \cdot 5H_2O$) can produce dimerized products that are high-density fuel candidates. There is one limitation of this reaction system, however, which involves forming not only the dimeric hydrocarbons that show molecular ion peaks at $m/z = 272$ $[C_{20}H_{32}]^+$ but also the oxygen-containing dimeric products, such as possibly dimeric alcohols or dimeric ethers, that show molecular ion peaks at $m/z = 290$ $[C_{20}H_{34}O]^+$. Although the inclusion of oxygen-containing species increases the overall density of the fuel mixture [5], which matches the purpose of augmenting the energy density of renewable fuel, this is not suitable considering the synthesis of hydrocarbons for drop-in fuel. Therefore, to exclude the production of oxygen-containing species from the reaction system catalyzed by $SnCl_4 \cdot 5H_2O$, we have started the rationalization of how such species are formed.

From the fact that the formation of bornyl chloride (3), one particular chlorine-containing side-product, the decomposition of Sn$^{(IV)}$ complex (A) during the reaction can be postulated (Scheme 2, left circle). In other words, to form compound 3, the protonation of α-pinene (1) by stannic complex A and successive ring enlargement by Wagner-Meerwein rearrangement occur first, furnishing a carbocation/Sn$^{(IV)}$ ate complex ion-pair (Int4). This intimate ion-pair Int4 then suffers from the dissociation of chloride ligand, producing both compound 3 and decomposed Sn$^{(IV)}$ complex (B).

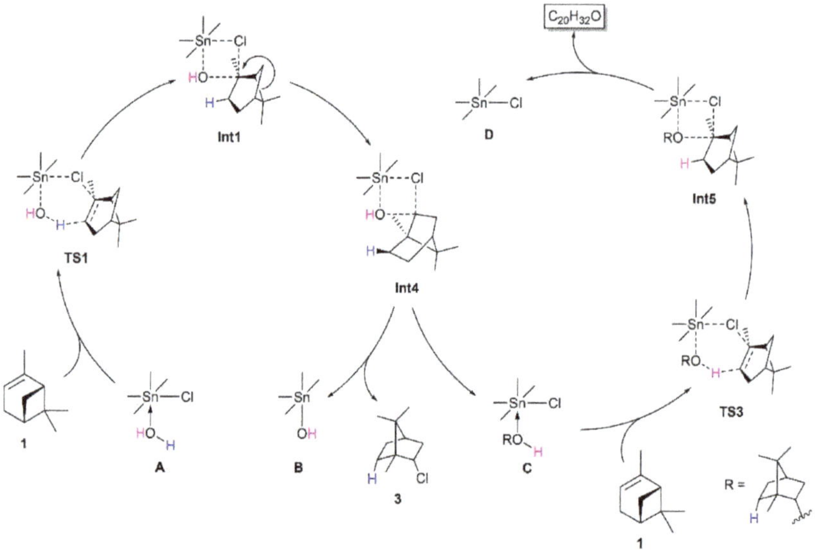

Scheme 2. A plausible mechanism of catalyst (**A**) decomposition suggested based on the production of bornyl chloride (**3**, left circle) and oxygen-containing dimeric products ($C_{12}H_{32}O$, right circle).

As previously mentioned, for the conversion of α-pinene (1) to occur, carbocation/Sn$^{(IV)}$ ate complex ion-pairs (Int1-4) should be reaction intermediates (Schemes 1 and 2). We considered that these intermediates are essential clues for the explanation of how the oxygen-containing species are formed. While bornyl chloride (3) is formed when the dissociation of chloride ligand (Sn-Cl bond) occurs in the intimate ion-pair Int4, monoterpene alcohols can be formed by the dissociation of hydroxo ligand (Sn-O bond). At this point, the monoterpene alcohols can occupy an empty coordinate site of the stannic center as an alcohol ligand, which generates ligand exchanged Sn$^{(IV)}$ complex (C) (Scheme 2, right circle). Since the same Lewis acid-assisted Brønsted acidic water ligands work as both catalysts and reactants, this alcohol ligand can also protonate another compound 1, furnishing a carbocation/Sn$^{(IV)}$ ate complex ion-pair (Int5) via TS3. From this intimate ion-pair Int5, the dissociation of the remained alkoxo ligand can cause the formation of dimeric ethers ($C_{20}H_{32}O$) and decomposed Sn$^{(IV)}$ complex (D).

Based on the above rationalization, we have concluded that, although the reaction system catalyzed by the Lewis acid-assisted Brønsted acid activity of $SnCl_4 \cdot 5H_2O$ can produce high-density fuel candidates from α-pinene, to synthesize dimeric products as only hydrocarbons, the water ligands must be switched. Instead of them, monodentate alcohols or phenol ligands, which will form alkoxo or aryloxo ligand after the protonation of α-pinene, respectively, also should be avoided since these ligands can furnish undesirable ether products. Therefore, we have considered bidentate 2-alkoxyphenol ligands such as 2-methoxyphenol (guaiacol) and 2,6-dimethoxyphenol (syringol), given they can be tightly bonded to the metallic center with their methoxy group even after the protonation of α-pinene [19–22]. To prove this hypothesis, further study should be followed.

4. Conclusions

To synthesize renewable high-density fuel from monoterpenes, we investigated whether the catalytic amount of stannic chloride molten salt hydrates ($SnCl_4 \cdot 5H_2O$) can furnish dimeric products from α-pinene in a solvent-less condition. $SnCl_4 \cdot 5H_2O$ would work as a catalyst for the dimerization of α-pinene under milder conditions than previously reported ones. This conversion was thought to be promoted by a Lewis acid-assisted Brønsted acidic nature of the coordinated water ligands on the stannic (Sn$^{(IV)}$) center. However, we can also observe the unexpected heteroatom-containing products of which oxygen and chlorine atoms resulted from the decomposition of the catalyst. Considering the proposed catalytic cycle and the mechanism of the catalyst decomposition, we suggested bidentate ligands to minimize the above problem.

Supplementary Materials: The following are available online at http://www.mdpi.com/2076-3417/10/21/7517/s1. Figure S1: GC/FID chromatograms of (A) the neat turpentine used in this study and (B) the reaction mixture after 1 h of reaction at 100 °C. Figure S2: Representative mass spectra of (A) dimeric hydrocarbons, (B) trimeric hydrocarbons, and (C) oxygen-containing dimeric products.

Author Contributions: Conceptualization, S.-M.C. Methodology, S.-M.C. and J.-H.C. Formal analysis, S.-M.C. and J.-H.K. Investigation, S.-M.C. and J.-H.C. Resources, S.-M.C. Data curation, J.-H.C. and J.-H.K. Writing—original draft preparation, S.-M.C. and J.-H.C. Writing—review and editing, S.-M.C. and B.K. Supervision, I.-G.C. All authors have read and agreed to the published version of the manuscript.

Funding: This study was supported by Mid-career Researcher Program in Basic Research of National Research Foundation of Korea grant funded by the Korea government (MSIP) (NRF-2016R1A2B4014222).

Conflicts of Interest: The authors declare no conflict of interest.

References

1. Xu, J.; Li, N.; Yang, X.; Li, G.; Wang, A.; Cong, Y.; Wang, X.; Zhang, T. Synthesis of diesel and jet fuel range alkanes with furfural and angelica lactone. *ACS Catal.* **2017**, *7*, 5880–5886. [CrossRef]
2. Xia, Q.; Xia, Y.; Xi, J.; Liu, X.; Zhang, Y.; Guo, Y.; Wang, Y. Selective One-Pot Production of High-Grade Diesel-Range Alkanes from Furfural and 2-Methylfuran over Pd/NbOPO$_4$. *ChemSusChem* **2017**, *10*, 747–753. [CrossRef] [PubMed]

3. Li, Z.; Otsuki, A.L.; Mascal, M. Production of cellulosic gasoline via levulinic ester self-condensation. *Green Chem.* **2018**, *20*, 3804–3808. [CrossRef]
4. Garcia, D.; Villa Holguín, A.L.; Lapuerta, M.; Bustamante, F.; Alarcón, E. Oxyfunctionalization of turpentine for fuel applications. *Energy Fuels* **2020**, *34*, 579–586. [CrossRef]
5. Cho, S.M.; Kim, J.H.; Kim, S.H.; Park, S.Y.; Kim, J.C.; Choi, I.G. A comparative study on the fuel properties of biodiesel from woody essential oil depending on terpene composition. *Fuel* **2018**, *218*, 375–384. [CrossRef]
6. Babu, A.M.; Saravanan, C.; Vikneswaran, M.; Jeo, V.E.; Sasikala, J. Visualization of in-cylinder combustion using endoscope in spark ignition engine fueled with pine oil blended gasoline. *Fuel* **2020**, *263*, 116707. [CrossRef]
7. Ashok, B.; Raj, R.T.K.; Nanthagopal, K.; Krishnan, R.; Subbarao, R. Lemon peel oil–A novel renewable alternative energy source for diesel engine. *Energy Convers. Manag.* **2017**, *139*, 110–121. [CrossRef]
8. Cho, S.M.; Hong, C.Y.; Park, S.Y.; Lee, D.S.; Choi, J.H.; Koo, B.; Choi, I.G. Application of Sulfated Tin (IV) Oxide Solid Superacid Catalyst to Partial Coupling Reaction of α-Pinene to Produce Less Viscous High-Density Fuel. *Energies* **2019**, *12*, 1905. [CrossRef]
9. Yuan, B.; Wang, Z.; Yue, X.; Yu, F.; Xie, C.; Yu, S. Biomass high energy density fuel transformed from α-pinene catalyzed by Brönsted-Lewis acidic heteropoly inorganic-organic salt. *Renew. Energy* **2018**, *123*, 218–226. [CrossRef]
10. Meylemans, H.A.; Baldwin, L.C.; Harvey, B.G. Low-temperature properties of renewable high-density fuel blends. *Energy Fuels* **2013**, *27*, 883–888. [CrossRef]
11. Deng, W.; Kennedy, J.R.; Tsilomelekis, G.; Zheng, W.; Nikolakis, V. Cellulose hydrolysis in acidified LiBr molten salt hydrate media. *Ind. Eng. Chem. Res.* **2015**, *54*, 5226–5236. [CrossRef]
12. Yang, X.; Li, N.; Lin, X.; Pan, X.; Zhou, Y. Selective cleavage of the aryl ether bonds in lignin for depolymerization by acidic lithium bromide molten salt hydrate under mild conditions. *J. Agric. Food Chem.* **2016**, *64*, 8379–8387. [CrossRef] [PubMed]
13. Yoo, C.G.; Zhang, S.; Pan, X. Effective conversion of biomass into bromomethylfurfural, furfural, and depolymerized lignin in lithium bromide molten salt hydrate of a biphasic system. *RSC Adv.* **2017**, *7*, 300–308. [CrossRef]
14. de Almeida, R.M.; Li, J.; Nederlof, C.; O'Connor, P.; Makkee, M.; Moulijn, J.A. Cellulose conversion to isosorbide in molten salt hydrate media. *ChemSusChem* **2010**, *3*, 325–328. [CrossRef] [PubMed]
15. Qiao, Y.; Pedersen, C.M.; Wang, Y.; Hou, X. NMR insights on the properties of $ZnCl_2$ molten salt hydrate medium through its interaction with $SnCl_4$ and fructose. *ACS Sustain. Chem. Eng.* **2014**, *2*, 2576–2581. [CrossRef]
16. Wang, Y.; Pedersen, C.M.; Qiao, Y.; Deng, T.; Shi, J.; Hou, X. In Situ NMR spectroscopy: Inulin biomass conversion in $ZnCl_2$ molten salt hydrate medium—$SnCl_4$ addition controls product distribution. *Carbohydr. Polym.* **2015**, *115*, 439–443. [CrossRef]
17. Hu, S.; Zhang, Z.; Song, J.; Zhou, Y.; Han, B. Efficient conversion of glucose into 5-hydroxymethylfurfural catalyzed by a common Lewis acid $SnCl_4$ in an ionic liquid. *Green Chem.* **2009**, *11*, 1746–1749. [CrossRef]
18. Omari, K.W.; Besaw, J.E.; Kerton, F.M. Hydrolysis of chitosan to yield levulinic acid and 5-hydroxymethylfurfural in water under microwave irradiation. *Green Chem.* **2012**, *14*, 1480–1487. [CrossRef]
19. Kumazawa, K.; Ishihara, K.; Yamamoto, H. Tin (IV) chloride-chiral pyrogallol derivatives as new Lewis acid-assisted chiral Brønsted acids for enantioselective polyene cyclization. *Org. Lett.* **2004**, *6*, 2551–2554. [CrossRef]
20. Nakamura, S.; Ishihara, K.; Yamamoto, H. Enantioselective biomimetic cyclization of isoprenoids using lewis acid-assisted chiral brønsted acids: Abnormal claisen rearrangements and successive cyclizations. *J. Am. Chem. Soc.* **2000**, *122*, 8131–8140. [CrossRef]
21. Ishihara, K.; Kaneeda, M.; Yamamoto, H. Lewis Acid assisted chiral Bronsted acid for enantioselective protonation of silyl enol ethers and ketene bis (trialkylsilyl) acetals. *J. Am. Chem. Soc.* **1994**, *116*, 11179–11180. [CrossRef]
22. Ishihara, K.; Nakashima, D.; Hiraiwa, Y.; Yamamoto, H. The crystallographic structure of a Lewis acid-assisted chiral Brønsted acid as an enantioselective protonation reagent for silyl enol ethers. *J. Am. Chem. Soc.* **2003**, *125*, 24–25. [CrossRef] [PubMed]
23. Nie, G.; Zou, J.-J.; Feng, R.; Zhang, X.; Wang, L. HPW/MCM-41 catalyzed isomerization and dimerization of pure pinene and crude turpentine. *Catal. Today* **2014**, *234*, 271–277. [CrossRef]

24. Lana, E.J.L.; da Silva Rocha, K.A.; Kozhevnikov, I.V.; Gusevskaya, E.V. One-pot synthesis of diisobornyl ether from camphene using heteropoly acid catalysts. *J. Mol. Catal. A Chem.* **2006**, *243*, 258–263. [CrossRef]

25. Sidorenko, A.Y.; Aho, A.; Ganbaatar, J.; Batsuren, D.; Utenkova, D.; Sen'kov, G.; Wärnå, J.; Murzin, D.Y.; Agabekov, V. Catalytic isomerization of α-pinene and 3-carene in the presence of modified layered aluminosilicates. *Mol. Catal.* **2017**, *443*, 193–202. [CrossRef]

26. Shihada, A.F.; Abushamleh, A.S.; Weller, F. Crystal Structures and Raman Spectra of cis-[SnCl$_4$ (H$_2$O)$_2$]·2H$_2$O, cis-[SnCl$_4$ (H$_2$O)$_2$]·3H$_2$O,[Sn$_2$Cl$_6$ (OH)$_2$ (H$_2$O)$_2$]·4H$_2$O, and [HL][SnCl$_5$(H$_2$O)]·2.5H$_2$O (L = 3-acetyl-5-benzyl-1-phenyl-4, 5-dihydro-1, 2, 4-triazine-6-one oxime, C$_{18}$H$_{18}$N$_4$O$_2$). *Z. Anorg. Allg. Chem.* **2004**, *630*, 841–847. [CrossRef]

27. Zhao, P.Q.; Xu, L.W.; Xia, C.G. Transition metal-based Lewis acid catalyzed ring opening of epoxides using amines under solvent-free conditions. *Synlett* **2004**, *2004*, 0846–0850. [CrossRef]

28. Wang, M.; Song, Z.; Liang, Y. SnCl$_4$·5H$_2$O-Catalyzed Synthesis of B-Amino Carbonyl Compounds Via a Direct Mannich-Type Reaction. *Prep. Biochem. Biotechnol.* **2010**, *41*, 1–6. [CrossRef]

29. Chen, D.; Wang, D.; Wu, W.; Xiao, L. SnCl$_4$·5H$_2$O: A Highly Efficient Catalyst for Hydration of Alkyne. *Appl. Sci.* **2015**, *5*, 114–121. [CrossRef]

30. Jasti, R.; Anderson, C.D.; Rychnovsky, S.D. Utilization of an oxonia-cope rearrangement as a mechanistic probe for prins cyclizations. *J. Am. Chem. Soc.* **2005**, *127*, 9939–9945. [CrossRef]

Publisher's Note: MDPI stays neutral with regard to jurisdictional claims in published maps and institutional affiliations.

 *applied sciences*

Review

Biorefinery: The Production of Isobutanol from Biomass Feedstocks

Yide Su, Weiwei Zhang *, Aili Zhang and Wenju Shao

School of Chemical Engineering and Technology, Hebei University of Technology, No. 8 Guangrong Road, Hongqiao District, Tianjin 300130, China; 201721508004@stu.hebut.edu.cn (Y.S.); zhangaili@tju.edu.cn (A.Z.); 201931505031@stu.hebut.edu.cn (W.S.)
* Correspondence: 201821508004@stu.hebut.edu.cn; Tel.: +86-22-60200444

Received: 16 October 2020; Accepted: 10 November 2020; Published: 20 November 2020

Abstract: Environmental issues have prompted the vigorous development of biorefineries that use agricultural waste and other biomass feedstock as raw materials. However, most current biorefinery products are cellulosic ethanol. There is an urgent need for biorefineries to expand into new bioproducts. Isobutanol is an important bulk chemical with properties that are close to gasoline, making it a very promising biofuel. The use of microorganisms to produce isobutanol has been extensively studied, but there is still a considerable gap to achieving the industrial production of isobutanol from biomass. This review summarizes current metabolic engineering strategies that have been applied to biomass isobutanol production and recent advances in the production of isobutanol from different biomass feedstocks.

Keywords: isobutanol; biorefinery; metabolic engineering; biomass utilization

1. Introduction

Energy and the environment are two major issues facing the world. Due to climate change and the demand for renewable transportation fuels, the production of environmentally friendly biofuels has aroused great interest. Compared with fossil fuels, biofuels are more sustainable and highly renewable, which has attracted much attention [1–7].

In the past few years, researchers have focused on the production of biofuels from edible crops [8]. This is not a long-term solution. From the perspective of economy and sustainability, the ultimate goal is to convert low-cost non-edible biomass resources into high-value biofuels and other chemical products [9–13]. Therefore, people have proposed the biotransformation of lignocellulosic biomass, most of which is agricultural waste. With the development of biorefinery, the potential industrialization of lignocellulosic biomass processing to release monosaccharides that can be fermented and converted into high-value chemicals has become a reality. Cellulosic isobutanol already occupies a place in the fuel sector [14].

Ethanol has obvious disadvantages compared to fossil fuels, but most current research in biorefinery is still based on ethanol production. We need to expand the types of chemicals that can be produced to facilitate further development of biorefinery. Compared with ethanol, most higher alcohols have lower hygroscopicity, higher energy density, higher octane number, and properties that are closer to gasoline, so they have a higher compatibility with existing equipment and higher operating safety factors [15–17]. Isobutanol is an important industrial compound. It is used in lubricants, coatings, adhesives, automobile spray paint, and as an intermediate for the synthesis of many drugs. In addition, isobutanol derivatives are widely used in the chemical industry [15]. As a new generation of biofuel, isobutanol has many advantages (Table 1) and uses [18–20]. Therefore, developing of biorefinery to produce biomass isobutanol is in line with market needs and sustainable development.

Table 1. Major characteristics comparison of several common biofuels with gasoline.

	Ethanol	1-butanol	Isobutanol	Gasoline
Lower Heating Value (MJ/kg)	27.0	33.1	33.3	43.5
Flash point (°C)	13	37	28	−43
Solubility (20 °C in water, wt %)	Miscible	7.7	8.7	negligible
Boiling temperature (°C)	78.4	117.7	108	25–215
Vapor toxicity	Toxic	Moderate	Moderate	Moderate

Biorefinery involves many steps of biomass treatment, biomass conversion, fermentation, product purification, fermentation processes, etc. In this article, we discuss the industrial microorganisms used in biorefinery, review the metabolic engineering strategies of isobutanol production by microorganisms, and introduce the current situation and developmental prospects of isobutanol production from various biomass feedstocks.

2. Research on Isobutanol Production

The isobutanol synthesis pathway (Figure 1) starts when pyruvate is converted to 2-ketoisovalerate (KIV) by acetolactate synthase (AHAS), acetohydroxyacid reductoisomerase (AHARI), and dihydroxyacid dehydratase (DHAD), respectively. Next, KIV is converted to isobutyraldehyde by the 2-ketoacid decarboxylase (KIVD), and finally alcohol dehydrogenase (ADH) produces isobutanol (we call the pathway consisting of these five enzymes the engineered isobutanol pathway below). Microorganisms that are widely used to biosynthesize isobutanol include *Escherichia coli*, *Corynebacterium glutamicum*, *Bacillus subtilis*, and *Saccharomyces cerevisiae*. Most microorganisms do not produce, or only produce trace amounts of isobutanol on their own. Now several strategies can increase isobutanol production, such as the overexpression of key enzymes of the isobutanol synthesis pathway, the inhibition of byproduct production, cofactor engineering, and microbial robustness enhancement. We selected some representative studies to briefly introduce current microbial isobutanol production strategies.

In 2008, Liao's group [16] validated the potential of microbes to produce higher alcohols using *E. coli*. By introducing KIVD from *Lactococcus lactis* and ADH from *S. cerevisiae*, the engineered strain JCL260 successfully produced isobutanol using glucose as a substrate. Then increasing the pyruvate and 2-ketoacid concentrations, JCL260 produced 22 g/L of isobutanol, which, at 86% of the theoretical maximum, demonstrated this strategy's potential. This is the first time that researchers have used metabolic engineering strategies to produce isobutanol with microorganisms. The cytotoxicity of isobutanol causes high concentrations of isobutanol to inhibit cell growth, thus limiting the maximum titer. Liao's group [21] further combined an in situ product removal strategy (gas stripping) with isobutanol production to improve the final titer of isobutanol. The final isobutanol production reached 50 g/L after 72 h of fed-batch fermentation in a bioreactor. They also used the chemical mutagen N'-nitro-N-nitrosoguanidine (NTG) to induced sequential mutagenesis in *E. coli* and screened for a high-yielding isobutanol strain. Repairing an inhibitory mutation resulted in a final isobutanol titer of 21.2 g/L isobutanol in 99 h [22]. Another factor that limits the isobutanol titer is a cofactor imbalance. Arnold's group [23] removed the dependence of the isobutanol synthesis pathway on NAPDH in *E. coli* and achieved 100% of the theoretical maximum isobutanol production during growth in anaerobic conditions.

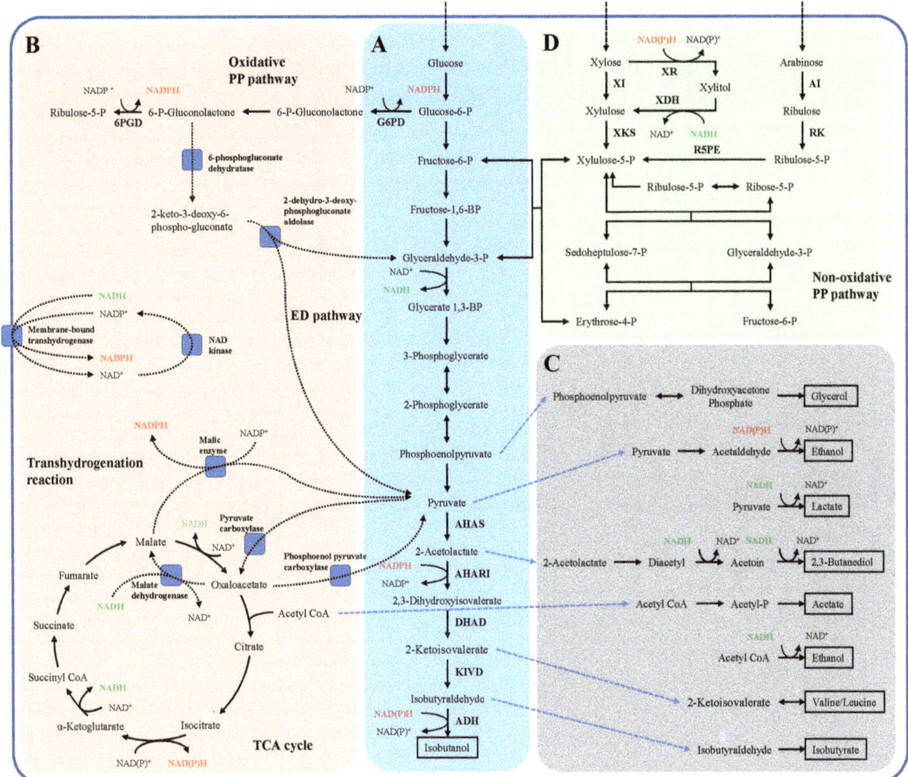

Figure 1. Schematic diagram of microorganism isobutanol synthesis pathways. Part (**A**) shows the general pathways of isobutanol biosynthesis; Part (**B**) shows a partial cofactor balancing strategy for successful optimization of isobutanol synthesis; Part (**C**) shows the common by-product synthesis pathways of isobutanol synthesis; and Part (**D**) shows the sugar metabolism pathway using non-glucose fermentation to produce isobutanol. The dashed line indicates the source of the precursor substance for the byproduct; AHAS: Acetolactate synthase; AHARI: Acetohydroxyacid reductoisomerase; DHAD: Dihydroxyacid dehydratase; KIVD: Keto acid decarboxylase; ADH: Alcohol dehydrogenase; G6PD: Glucose-6-phosphate dehydrogenase; 6PGD: 6-phosphogluconate dehydrogenase; XR: Xylose reductase; XDH: Xylitol dehydrogenase; XI: Xylose isomerase; XKS: Xylulokinase; AI: L-arabinose isomerase; RK: L-Ribulokinase; R5PE: L-Ribulose-5-P-4-epimerase.

C. glutamicum is widely used to produce various amino acids, so some researchers have explored isobutanol production by *C. glutamicum*. In 2010, Liao's group [24] tried to produce isobutanol with *C. glutamicum* for the first time. They overexpressed the engineered isobutanol pathway, and on this basis deleted the *PYC* gene (encoding pyruvate carboxylase) and *LDH* gene (encoding lactate dehydrogenase). This engineered strain they obtained produced 4.9 g/L of isobutanol when fermented for 96 h. Blombach's group [25] found that the main byproducts of *C. glutamicum* during isobutanol production are lactic and succinate. By knocking out the byproduct synthesis genes, restoring the redox balance in combination with heterologous expression of transhydrogenase (*pntAB* from *E. coli*), and overexpressing *adhA*, the final isobutanol titer reached 13 g/L at 48 h. Recently, Inui's group [26] has used different promoter combinations to confirm the importance of higher activity of AHAS and KDC for isobutanol synthesis in *C. glutamicum*. In combination with a cofactor strategy (altering the cofactor specificity of AHARI and ADH) and enhanced glycolytic flux strategy (overexpression of endogenous glycolytic genes and the phosphoenolpyruvate:carbohydrate phosphotransferase system

(PTS), plus introduction of the Entner–Doudoroff pathway from *Zymomonas mobilis*), 20.8 g/L isobutanol was produced at 24 h to reach 84% of the theoretical value.

Wen's group [27] proved that *B. subtilis* could be used as a cell factory for isobutanol synthesis through an isobutanol tolerance test. In 2011, they introduced the engineered isobutanol pathway into *B. subtilis*, resulting in strain BSUL03. The concentration of the isobutanol reached 2.63 g/L at 54 h. Next, they performed an elementary mode analysis (EMA) on the engineered strain BSUL03 to identify targets in the metabolic network that could be optimized, lactate dehydrogenase and pyruvate dehydrogenase complexes. By knocking out the *ldh* and *pdhC* genes, they engineered the strain BSUL05 and obtained an isobutanol titer of 5.5 g/L after 60 h of fermentation [28]. Afterwards, they performed a metabolic flux analysis and comparison of the two engineered strains. To increase the concentration of NADPH and achieve a redox balance, the engineered strain BSUL08 was obtained by knocking out *pgi* (encoding glucose 6-phosphate isomerase), overexpressing *zwf* (encoding glucose 6-phosphate dehydrogenase), and further overexpression the transhydrogenase gene (*udhA* from *E. coli*). The production performance was tested with fed-batch fermentation and the isobutanol titer reached 6.12 g/L at 60 h, which was 63% of the theoretical maximum [29].

S. cerevisiae can produce a small amount of higher alcohols through the biosynthetic pathway of various amino acids. It also has a natural tolerance to alcohols that is higher than other microorganisms [30]. In recent years, *S. cerevisiae* has been widely used to produce higher alcohols. In 2011, Chen's group [31] overexpressed of the genes *ILV2* (encoding acetolactate synthase), *ILV3* (encoding dihydroxy acid dehydratase), and *ILV5* (encoding diacetolactate reductase) to obtain an isobutanol yield of 3.86 mg/g glucose. This is the first record of *S. cerevisiae* being used to produce isobutanol. Based on that, Boles's group [32] truncated the N-terminal mitochondrial targeting sequence of the *ILV*s to achieve expression of *ILV2*, *ILV3*, and *ILV5* in the cytoplasm, and optimized the codons of these three genes. They studied the activities of KIVD and ADH and determined that *ARO10* and *ADH2* were the most active enzymes in the synthesis of isobutanol. The final concentration of isobutanol produced by the strain they obtained was 0.63 g/L. Then they [33] knocked out the synthesis pathway of leucine, isoleucine, 2,3-butanediol, glycerol, pantothenate, and isobutyrate, thereby increasing the carbon flux of the isobutanol metabolism pathway. Here, the isobutanol titer reached 2.09 g/L in 96 h. Ethanol has always been the largest byproduct of isobutanol production, but the complete removal of pyruvate decarboxylase activity would also arrest cell growth. Therefore, Avalos's group [34] designed two powerful optogenetic gene expression systems in *S. cerevisiae*. In the presence of light, this system induces the expression of *PDCs* and promotes cell growth; in the dark, it induces the expression of *ILV2* to enhance the isobutanol biosynthesis. By controlling light exposure during fermentation, they could control ethanol production and could increase the isobutanol titer to 8.49 g/L.

3. Biomass Isobutanol Production

3.1. Isobutanol Production from Lignocellulose

Isobutanol is considered a promising alternative to gasoline, and cellulosic isobutanol is becoming increasingly important in the wave of next-generation biofuels. Extensive research in cellulosic ethanol [35,36] and cellulosic butanol [37,38] has provided a theoretical basis sufficient for the synthesis of cellulosic isobutanol. Some researchers have carried out feasibility analyses on the industrial production of cellulose isobutanol [39,40], confirming the great potential of cellulose isobutanol. However, few studies focus on the synthesis of isobutanol from lignocellulose biomass (details are summarized in Table 2).

3.1.1. Cellulosic Isobutanol Produced by Natural Cellulose-Degrading Microorganisms

Degrading cellulose with microorganisms that naturally utilize lignocellulose and convert the resulting C5 and C6 sugars into isobutanol is the simplest way to obtain cellulosic isobutanol. To minimize the lignocellulose glycation process and production costs, the consolidated bioprocessing

(CBP) strategy was developed. The first cellulosic isobutanol was synthesized by introducing the engineered isobutanol pathway into *Clostridium cellulolyticum* [41], which utilizes cellulose naturally. The strain produced a final isobutanol titer of 0.66 g/L. The simple introduction of the isobutanol synthesis pathway resulted in very low amounts of isobutanol titer, which is not sufficient for industrial scale production. The inability of *C. cellulolyticum* to process large amounts of substrate has been reported [42]. Destroying the ability of *C. cellulolyticum* to form spores by knocking out the *spo0A* gene improved cellulose utilization, but there was little variation in isobutanol titer (even lower than the wild type when the cellulose substrate reaches 50 g/L) [43]. The biosynthesis of cellulose isobutanol was also attempted in *Geobacillus thermoglucosidasius* [44], which produced isobutanol at 0.6 g/L from cellobiose in 60 h. While unsuccessful at producing large volumes of cellulosic isobutanol, this experiment demonstrated the strong thermal stability of ALAS and KIVD, which are widely used for isobutanol synthesis.

Table 2. Summary of microbial utilization of biomass to produce isobutanol.

Microorganism	Carbon Source	Strategy	Genes Involved	Titer	Time	Reactor	Reference
Clostridium cellulolyticum	Cellulose	Engineered isobutanol pathway	$ilvD_{EC}$, $ilvC_{EC}$, $yqhD_{EC}$, $alsS_{BS}$, $kivd_{LL}$	0.66 g/L	216 h	Tube	[41]
	Cellulose	Keto acid pathway Promoter engineering	$\Delta spo0A$, $alsS_{BS}$, $kivd_{LL}$	0.35 g/L	~250 h	Unknown	[43]
Geobacillus thermoglucosidasius	Cellobiose	Keto acid pathway Promoter engineering	$ilvC_{GT}$ $alsS_{BS}$, $kivd_{LL}$ (LLKF_1386)	0.6 g/L	48 h	Tube	[44]
Clostridium thermocellum	Cellulose	Keto acid pathway Promoter engineering Optimize fermentation conditions	$ilvB_{CT}$, $ilvN_{CT}$, $ilvC_{CT}$, $ilvD_{CT}$, $kivd_{LL}$	5.4 g/L	75 h	Tube	[45]
	Cellulose	Inhibition competition pathway Adaptive laboratory evolution	Δhpt, Δldh, Δpta, $adhE^{D494G}$	5.1 g/L	220 h	Bioreactor	[46]
Trichoderma reesei and *Escherichia coli*	Pretreated corn stover	Random mutagenesis Engineered isobutanol pathway Microbial consortium	*T. reesei* RUTC30: - *E. coli* NV3: $ilvC_{EC}$, $ilvD_{EC}$, $alsS_{BS}$, $kivd_{LL}$, $Adh2_{SC}$	1.88 g/L	380 h	Bioreactor	[47]
Caldicellulosiruptor bescii	Switchgrass	Inhibition competition pathway AOR-ADH pathway	Δldh $PF0346_{PF}$ (AOR), $Teth514\_0564_{PF}$ (ADHA)	0.17 g/L	40 h	Fermentor	[48]
	Glucose-xylose mixture	Dismantle carbon catabolite repression Inhibition competition pathway Engineered isobutanol pathway	$\Delta ldhA$, $\Delta adhE$, $\Delta pflB$, $\Delta pta\text{-}ackA$, mlc^*, $ilvC_{EC}$, $ilvD_{EC}$, $alsS_{BS}$, $kivd_{LL}$, $Adh2_{SC}$,	11 g/L	182 h	Flask	[49]
	Cedar	Dismantle carbon catabolite repression Inhibition competition pathway Promoter engineering Chromosome integration Optimize fermentation conditions	$\Delta ldhA$, $\Delta adhE$, $\Delta pflB$, $\Delta pta\text{-}ackA$, mlc^*, $ilvC_{EC}$, $ilvD_{EC}$, $alsS_{BS}$, $kivd_{LL}$, $adhA_{LL}$	3.7 g/L	96 h	Flask	[49,50]

Table 2. *Cont.*

Microorganism	Carbon Source	Strategy	Genes Involved	Titer	Time	Reactor	Reference
Saccharomyces cerevisiae	Xylose	Xylose XI pathway Cytosolic isobutanol pathway	$\Delta Ilv2$, $\Delta Ilv5$, $\Delta Ilv3$, $xylA_{CP}$, $Tal1_{SC}$, $Xks1_{SC}$, $Ilv2\Delta N54_{SC}$, $Ilv5\Delta N48_{SC}$, $Ilv3\Delta N19_{SC}$, $Aro10_{SC}$, $Adh2_{SC}$	1.36 mg/L	150 h	Flask	[51]
	Xylose	Xylose XI pathway Chromosome integration Adaptive laboratory evolution Mitochondrial isobutanol pathway Fed-batch fermentation	$\Delta BAT1$, $\Delta ALD6$, $\Delta PHO13$, $\Delta URA3$ $RKI1_{SC}$, $RPE1_{SC}$, $TKL1_{SC}$, $TAL1_{SS}$, $XYLA_{PE}$, $XYL3_{SS}$, $ILV2_{SC}$, $ILV5_{SC}$, $ILV3_{SC}$, $kivd_{LL}$, $AdhA_{LL}^{RE1}$	3.1 g/L	192 h	Tube	[52,53]
	Xylose	Xylose XR-XDH pathway Chromosome integration Mitochondrial isobutanol pathway Copy number optimization Adaptive laboratory evolution	$\Delta PHO13$, $\Delta GRE3$, $hxt7^{F79S}$, $XYL1_{SS}$, $XYL2_{SS}$, $XYL3_{SS}$ $ILV2_{SC}$, $ILV5_{SC}$, $ILV3_{SC}$, $ADH7_{SC}$, $kivd_{LL}$	92.9 mg/L	144 h	Tube	[53,54]
	Xylose	Xylose XR-XDH pathway Chromosome integration Copy number optimization Mitochondrial isobutanol pathway Optimize fermentation conditions Fed-batch fermentation	$\Delta ALD6$, $\Delta PHO13$ $XYL1_{SS}$, $XYL2_{SS}$, $XYL3_{SS}$ $ILV2_{SC}$, $ILV5_{SC}$, $ILV3_{SC}$, $kivd_{LL}$, $AdhA_{LL}^{RE1}$	2.6 g/L	Unknown	Bioreactor	[55,56]
Corynebacterium glutamicum	Hemicellulose fraction	Inhibition competition pathway Xylose XI pathway Arabinose metabolism pathway Engineered isobutanol pathway	Δpqo, $\Delta ilvE$, $\Delta ldhA$, Δmdh, $xylA_{XC}$, $xylB_{CG}$, $araB_{EC}$, $araA_{EC}$, $araD_{EC}$, $ilvB_{EC}$, $ilvN_{EC}$, $ilvC_{EC}$, $ilvD_{EC}$, $pntAB_{EC}$, $kivd_{LL}$, $Adh2_{CG}$	0.53 g/L	~28 h	Flask	[57,58]

Table 2. *Cont.*

Microorganism	Carbon Source	Strategy	Genes Involved	Titer	Time	Reactor	Reference
	Cellobiose	Inhibition competition pathway Copy number optimization Engineered isobutanol pathway Cellobiose metabolism pathway	$\Delta adhE$, $\Delta frdBC$, Δfnr, $\Delta ldhA$, Δpta, $\Delta pflB$, $ilvC_{EC}$, $ilvD_{EC}$, $alsS_{BS}$, $kivd_{LL}$, $adhA_{LL}$, $bglC_{TF}$	7.64 g/L	72 h	Unknown	[59]
	Cellobionic	Inhibition competition pathway Engineered isobutanol pathway	$\Delta adhE$, $\Delta frdBC$, Δfnr, $\Delta ldhA$, Δpta, $\Delta pflB$, $ilvC_{EC}$, $ilvD_{EC}$, $alsS_{BS}$, $kivd_{LL}$, $adhA_{LL}$	1.4 g/L	48 h	Unknown	[60]
Corynebacterium crenatum	Duckweed	Engineered isobutanol pathway	$ILV2_{SC}$, $ILV5_{SC}$, $ILV3_{SC}$, $kivd_{LL}$, $Adh2_{SC}$	1.15 g/L	96 h	Flask	[61]
	Duckweed	Whole-cell mutagenesis Engineered isobutanol pathway Simultaneous saccharification and fermentation	$ILV2^*_{SC}$, $ILV5^*_{SC}$, $ILV3_{SC}$, $kivd_{LL}$, $Adh6^*_{SC}$	5.6 g/L	96 h	Flask	[62]
	Emptyfruit bunches	Engineered isobutanol pathway Optimize fermentation conditions Separate hydrolysis and fermentation	$ilvC_{EC}$, $ilvD_{EC}$, $adhP_{EC}$, $alsS_{BS}$, $kivD_{LL}$	5.4 g/L	156 h	Unknown	[62]
Enterobacter aerogenes	Sugarcane bagasse	Inhibition competition pathway Engineered isobutanol pathway Pervaporation-coupled fermentation	$\Delta ldhA$, $\Delta budA$, $\Delta pflB$, $\Delta ptsG$, $ilvD_{KP}$, $ilvC_{KP}$, $budB_{KP}$, $kivD_{LL}$, $adhA_{LL}$	23 g/L	72 h	Fermenter	[63,64]
Escherichia coli	Algal protein	Chemical mutagenesis Protein conversion Cofactor engineering	$\Delta glnA$, $\Delta gdhA$, $\Delta luxS$, $\Delta lsrA$, $ilvC^{A71S, R76D, S78D, Q110A}$, $yqhD^{G39I, S40R}$ $ilvE_{EC}$, $ilvA_{EC}$, $sdab_{EC}$, $avta_{EC}$, $LueDH_{TI}$ $ilvD_{EC}$, $alsS_{BS}$, $kivD_{LL}$	0.2 g/L	Unknown	Flask	[65,66]

Table 2. *Cont.*

Microorganism	Carbon Source	Strategy	Genes Involved	Titer	Time	Reactor	Reference
E. coli BLF2 and *E. coli* AY3 (1:1.5)	Distillers' grains	Chemical mutagenesis Protein conversion Cofactor engineering Engineered isobutanol pathway Microbial consortium	*E. coli* BLF2: Δ*ldh* *ilvC*$_{EC}$, *ilvD*$_{EC}$, *YqhD*$_{EC}$, *alsS*$_{BS}$, *kivd*$_{LL}$ *E. coli* AY3: Δ*glnA*, Δ*gdhA*, Δ*luxS*, Δ*lsrA*, *ilvC*$^{A71S, R76D, S78D, Q110A}$, *yqhD*$^{G391, S40R}$ *ilvE*$_{EC}$, *ilvA*$_{EC}$, *sdab*$_{EC}$, *aota*$_{EC}$, *LueDH*$_{TI}$ *ilvD*$_{EC}$, *alsS*$_{BS}$, *kivD*$_{LL}$	6.5 g/L	52 h	Tube	[65,67]
Bacillus subtilis	Okara wastes	Activation of ilv-leu operonInhibition competition pathwayKeto acid pathway	Δ*codY*, Δ*bkdB*, Δ*relA*,*LueDH*$_{TI}$,*kivD*$_{LL}$, *yqhD*$_{EC}$	0.02 g/L	Unknown	Flask	[68]
Bacillus subtilis and *Escherichia coli* (1:4)	Watermelon rind and Okara waste	Protein conversion Engineered isobutanol pathway Microbial consortium	*B. subtilis*: Δ*codY*, Δ*bkdB*, *LueDH*$_{TI}$, *kivD*$_{LL}$, *yqhD*$_{EC}$ *E. coli*: *ilvC*$_{EC}$, *ilvD*$_{EC}$, *YqhD*$_{EC}$, *alsS*$_{BS}$, *kivd*$_{LL}$	0.88 g/L	220 h	Flask	[69–72]

Note: The abbreviation in the upper left corner of the gene indicates that the gene is a mutation; abbreviations in the lower right corner of genes indicate microorganisms of gene origin. BS: *Bacillus subtilis*; CG: *Corynebacterium glutamicum*; CP: *Clostridium phytofermentans*; CT: *Clostridium thermocellum*; EC: *Escherichia coli*; GT: *Geobacillus thermoglucosidasius*; KP: *K. pneumoniae* KCTC2242; LL: *Lactococcus lactis*; PE: *Piromyces* sp. E2; PF: *Pyrococcus furiosus*; PS: *Pichia stipites*; SC: *Saccharomyces cerevisiae*; SS: *Scheffersomyces stipitis*; TF: *Thermobifida fusca*; TI: *Thermoactinomycetes intermedius*; XC: *Xanthomonas campestris*.

The cellulose-utilizing *Clostridium thermocellum* is considered a promising producer of cellulosic biofuels, so additional experiments have attempted cellulosic isobutanol synthesis in this bacterial species [69]. Refinement gene expression systems [70–72] and gene editing techniques [73] have already been developed for *C. thermocellum*, and its basic metabolic network has been explored. Introducing the KIVD into *C. thermocellum*, optimizing the isobutanol synthesis pathway, and limiting the urea content of the medium enabled the engineered strain to ferment 5.4 g/L of isobutanol in 75 h [45]. Experiments have also identified a new pathway in *C. thermocellum* that converts KIV to isobutanol. This pathway decarboxylates KIV to isobutyryl-CoA with a ketoisovalerate ferrooxide-dependent reductase (KOR), which is converted to isobutanol by the aldehyde/alcohol dehydrogenase (ADH). Most recently, Holwerda's group [46] engineered *C. thermocellum* by eliminating the acetic and lactic synthesis pathways and performed adaptive laboratory evolution, resulting in the strain LL1043 that produced an isobutanol titer of 5.1 g/L. High titer were obtained without optimizing isobutanol synthesis, suggesting that the production of cellulose isobutanol with *C. thermocellum* remains a possibility.

Some new studies attempt to apply isobutanol production to other microbes. Lin's group [47] designed a microbial consortium that cocultured the cellulose utilizing strain *Trichoderma reesei* with an isobutanol-producing strain of *E. coli* to produce 1.88 g/L of isobutanol from pretreated corn stover. Another study [48] investigated a new AOR-ADH pathway for alcohol production. This pathway converts acetate to ethanol via aldehyde ferredoxin oxidoreductase (AOR) and ADH. The cellulolytic extreme thermophile *Caldicellulosiruptor bescii* naturally produces acetate. Heterologous expression of the AOR-ADH pathway in *C. bescii*, resulted in ethanol synthesis. Isobutanol was synthesized using switchgrass as a carbon source following isobutyrate supplementation, providing new ideas for subsequent isobutanol synthesis.

3.1.2. Cellulosic Isobutanol Produced by Non-Native Cellulose-Degrading Microorganisms

Another strategy for obtaining cellulosic isobutanol is to further reform the pre-existing engineered strains for isobutanol production to utilize lignocellulose or its pretreatment products. *E. coli* and *S. cerevisiae* are the most promising cellulosic isobutanol-producing strains. Two model organisms have invested considerable research in cellulose utilization, which have been summarized in a considerable number of reviews [74,75]. They both have significant carbon catabolite repression and cannot utilize glucose simultaneously with other sugars. Xylose is the most abundant sugar in lignocellulose aside from glucose, so many efforts have been to construct industrially viable xylose-utilizing strains. To eliminate carbon catabolite repression, *E. coli* used UV mutagenized screened a new target gene [49], *Mlc*, encoding a DNA-binding transcriptional repressor. A strain with this mutant *Mlc* gene outperformed previous research in fermenting isobutanol from mixed glucose and xylose. Further optimization of the isobutanol synthesis pathway genes expression (*ilvC*, *ilvD*, *alsS*, *kivd* and *adhA*) with the biomass-inducible chromosome-based expression system (BICES) [50], achieved an isobutanol titer of 3.7 g/L from cedar hydrolysate.

S. cerevisiae cannot convert xylose naturally, but considerable efforts have been made to engineer a strain of *S. cerevisiae* that can [76]. There are two types of xylose metabolism (Figure 1D), one catalyzed by xylose reductase (XR) and xylitol dehydrogenase (XDH), the other by xylose isomerase (XI). However, the XR-XDH pathway requires the cofactor NADPH that is also in high demand for isobutanol synthesis, prone to cofactor imbalance. For this reason, the XI pathway is favored for isobutanol synthesis, but most attempts to express heterologous xylose isomerases in *S. cerevisiae* have failed. XylA, a xylose isomerase from *Clostridium phytofermentans*, was successfully expressed in *S. cerevisiae* by Boles' group [51]. Further, they increased xylose metabolism by overexpressing *Xks1* and *Tal1*, with the redesigned cytoplasmic isobutanol pathway, it achieved first isobutanol production by using xylose as the sole carbon source. Stephanopoulos' group engineered a strain to produce ethanol from xylose by the introducing xylose isomerase from *Piromyces sp.* E2 in *S. cerevisiae*. Overexpressing *RKI1*, *RPE1*, *TKL1*, *TAL1* and *XYL3* in this background increased xylose assimilation and facilitated adaptive laboratory evolution [52]. Boles' group [53] further optimized the use of xylose in this strain by knocking out gene *PHO13*. With expression of the mitochondrial isobutanol pathway and inhibition of valine and acetic acid synthesis, the production of isobutanol was 3.1 g/L in 192 h of fed-batch fermentation. These experiments show that xylose promotes mitochondrial activity significantly more than glucose does and increases with increasing concentrations of xylose. Thus, the advantages of isobutanol production through the mitochondrial isobutanol pathway from xylose were demonstrated. Another xylose utilization pathway, the XR-XDH pathway, has also been used to produce isobutanol. Runguphan's group [77] determined through combinatorial screening that the xylose metabolism pathway from *Scheffersomyces stipitis* is most effective when applied to isobutanol synthesis. Incorporating the strategies identified in previous studies to enhance xylose conversion and assimilation (knockout *PHO13*, *GRE3* and overexpression of *XYL3*) and optimizing the copy number of isobutanol pathway genes produced 48.4 mg/L of isobutanol within 144 h. Adaptive laboratory evolution of the resulting engineered strain, and identification of two newly discovered mutation targets (the $CCR4^{A638S}$ and TIF^{A79S}) to improve xylose utilization [54] increased the xylose ratio growth rate by 40.6%, but had little impact on isobutanol production. Using the XR-XDH pathway is likely to cause redox imbalance, and numerous studies have attempted to alleviate the cofactor imbalance by altering the cofactor specificity of the enzyme [78–80]. Jin's group [55] adjusted the copy numbers of three genes (*XYL1*, *XYL2*, and *XYL3*) in *S. cerevisiae* to improve redox balance and reduce acetate and xylitol accumulation. These results of evolutionary engineering confirmed the importance of deleting *PHO13* for xylose utilization. The final isobutanol titer was 2.6 g/L when coupled with the mitochondrial isobutanol pathway [56]. Metabolite analysis of the engineered strain revealed that the use of xylose increased valine levels in *S. cerevisiae*, which can be converted by branched-chain amino acid transaminases into KIV. This once again confirms that xylose promotes isobutanol synthesis.

C. glutamicum is also a promising industrial producer of isobutanol, and it is necessary to develop an engineered strain that can utilize cellulose. By heterologous expression the xylose XI and the arabinose metabolic pathways from *E. coli*, Blombach's group [57] succeeded in constructing strains that can rapidly utilize mixed sugars, including glucose. When combined with previous strategies to optimize isobutanol synthesis [25,58], 0.53 g/L of isobutanol could be produced from hemicellulose fractions.

One way to address the fermentation of mixed sugars containing xylose is to bypass the use of glucose. Cellobiose is an intermediate product of cellulose hydrolysis to glucose and has no carbon catabolite repression effects on cells. Mixed sugar fermentation using cellobiose and xylose has been reported [81,82], but no studies have addressed isobutanol synthesis. Isobutanol production using cellobiose was attempted based on a previously engineered strain from *E. coli*. Expression of β-glucosidase from *Thermobifida fusca* resulted in an isobutanol production of 7.64 g/L by optimizing gene copy number [59]. It should be noted that direct transport of cellobiose to intracellular hydrolysis is required to avoid carbon catabolite repression, and no growth of cells expressing the cytoplasmic enzyme was observed in this study. Another study [60] tested cellobionic acid, the main product of cellulose hydrolysis assisted by the use of lytic polysaccharide monooxygenase (LPMO), as a carbon source for the production of isobutanol. *E. coli* was found a naturally utilize the cellobionic acid pathway. A final titer of 1.4 g/L was produced from the cellulose hydrolysate of *Neurospora crassa* conversion, thus expanding the available carbon source for cellulosic isobutanol.

Most of the studies described above are based on either industrially produced cellulose or the monosaccharides obtained from its hydrolysis as substrates, but there is a lack of studies on the direct use of cellulose or its hydrolysates to produce isobutanol. Depending on the source of cellulosic biomass, the proportions of cellulose, hemicellulose, and lignin can vary considerably [83]. The hydrolysis products will also be significantly different, depending on how the pretreatment is metabolized [84]. These hydrolysis products (weak acids, furan derivatives, phenolic compounds, etc.) toxic to cells and reduce the isobutanol titer. Several experiments have tested the ability of engineered isobutanol-producing strains designed to ferment the hydrolysis products of cellulosic biomass. Duckweed [61,85] is a fast-growing non-food crop with few growth requirements that is very easy to handle. Empty fruit bunches [62] are a large byproduct of palm oil production and sugarcane bagasse [63] is a common industrial byproduct. Their pretreated fermentation inhibitor content is low, as can be concluded from the comparison with glucose fermentation. This resulted in these biomasses being good carbon source providers. The best isobutanol titer was obtained from *Enterobacter aerogenes* [64] use of sugarcane bagasse hydrolysates to produce 23 g/L of isobutanol by pervaporation-coupled fermentation.

3.2. Isobutanol Production from Protein

Biorefinery is focused on the industrial production of high-value compounds from biomass. Industrial waste like distillers' grains and okara waste cannot be integrated into the most existing strategies due to their primary component, proteins. For the process of protein hydrolysis to amino acids, some amino acids produce 2-keto acids, which are precursors of isobutanol. Liao's group [65,86] has pioneered the use of proteins as feedstock for higher alcohols production. In *E. coli*, they used chemical mutagenesis to screen the *YH19* strain that can utilize 13 amino acids as a sole carbon source. The restriction of ammonia assimilation, introduction of three exogenous transamines and deamines cycles, and knocking out population-sensing genes (*luxS* and *lsrA*) generated the strain YH83 [65]. Strain YH83 expresses the engineered isobutanol pathway that used algal protein hydrolysates to produce higher alcohols. Taking the more severe cofactor imbalance triggered by the use of protein biomass into consideration, Davis' group [66] performed cofactor-specific reconstructions of 2,3-dihydroxy isovalerate oxidoreductase (IlvC) and NADPH-dependent isobutanal dehydrogenase (YqhD) in the *YH83* strain to obtain an isobutanol titer of 0.2 g/L from algae protein hydrolysate. Davis' group [67] then attempted to utilize both sugars and proteins from biomass by constructing an *E. coli* microbial consortium, that is coculturing a proportional mix of strain *AY3*, which can produce heteroalcohols from proteins, and strain *BL2*, which can produce isobutanol from glucose and xylose. Protein utilization increased from 16.3% to 31.3% in a separate culture at a ratio of 1.5:1. The fermentation of distillers' grains hydrolysate under these conditions also resulted in the highest isobutanol titer of 6.5 g/L, which was higher than the 5.5 g/L achieved with strain *BL2* fermentation alone. The fermentation of other biomass like algal protein hydrolysates was also investigated and the highest isobutanol titer of 2.38 g/L was achieved when strains *AY3* and *BL2* were fermented in a 4:1 ratio, demonstrating the industrial potential of this microbial consortium.

Since *E. coli* does not naturally produce the proteases required to process protein biomass, further attempts were made in *B. subtilis*, which secretes proteases that enable growth on polypeptides. Due to the different genetic backgrounds of *E. coli* and *B. subtilis*, a series of experiments were performed to determine the importance of codY (a global regulator) deletion for inhibiting ammonia assimilation and improving branched-chain amino acid synthesis. Combined with a strategy of preventing the degradation of branched-chain amino acid by knocking out the dihydrolipoyl acyltransferase (*bkdb*) in the branched-chain 2-keto acid dehydrogenase complex, a final biofuel titer of 0.72 g/L was obtained from protein biomass [86]. Based on this, Choi's group [68] knocked out *RelA*, expressing the regulatory protein responsible, that recognizes nutritional stress to activate the ilv-leu operon [87] and further promotes branched-chain amino acids synthesis. Comparing the results of spent coffee grounds and okara wastes fermentation, the triple deletion strain expressing KIVD and ADH produced almost no isobutanol with spent coffee grounds hydrolysate as a substrate, while okara wastes hydrolysate fermented about 0.02 g/L of isobutanol in eight days. The microbial consortium strategy was also applied to this engineered strain and isobutanol was successfully obtained from the hydrolysates of watermelon rind and soybean residue by coculturing with *E. coli AY3* [88]. The highest isobutanol titer was 0.88 g/L at an *E. coli* to *Bacillus subtilis* ratio of 4:1. From these studies that used proteins to produce isobutanol, it was found that differences in the source of biomass could lead to large changes in product ratios, which require further research.

4. Conclusions

In recent years, researchers have made considerable effort to increase the yield of microbial biofuel production and continued to explore chemical synthesis from different biomass materials [13]. In addition, some achievements have been made in the production of bioethanol [14,89]. In this paper we summarized current metabolic engineering strategies applied to biomass isobutanol production and recent advances in the production of isobutanol from different biomass feedstocks.

It can be seen that in the last decade, there was a significant improvement in the production of cellulosic isobutanol while the synthesis of isobutanol from protein biomass was demonstrated. Biomass isobutanol is no longer at the theoretical stage. However, the efforts made in the broader context of biorefining were not enough. Since isobutanol is a secondary metabolite, controlling the competing pathway to maximize the conversion of pyruvate to isobutanol is difficult, and removing the largest competitor, the ethanol pathway, can lead to severe growth defects. Coupled with the fact that biomass utilization efficiencies are low, the combination of the two problems leads to very low yields. Efficient inhibition of ethanol synthesis has been accomplished using light-controlled genetic systems [34], but the in vivo metabolic backgrounds of different microorganisms are very different, and further investment in research is needed to accomplish an efficient inhibition of the competing pathways. Another issue is how to enhance the robustness of engineered microorganisms, as both isobutanol and organics from biomass hydrolysates can cause considerable cellular damage. This issue includes how to improve the redox imbalance caused by isobutanol production, as it has been shown an improved redox significantly increases strain isobutanol tolerance [90] or furfural tolerance [91–93] and thus isobutanol production. Adaptive laboratory evolutionary applications for tolerance enhancement have obtained some results, with a large number of target genes being identified. However, the tolerance mechanisms are still opaque for both isobutanol and lignocellulose-derived microbial inhibitors. Research on the application of tolerance enhancement strategies to biomass isobutanol production is also quite scarce.

In particular, it should be noted that of all the current studies on the production of isobutanol from biomass, the highest titers reach 23 g/L by pervaporation-coupled fermentation, while the highest titers of isobutanol production are currently over 50 g/L due to in situ product removal. The timely isolation of isobutanol from the medium reduces the inhibitory effect of isobutanol on cells, reduces cytotoxicity, and significantly enhances the final titer. Fermentation processes also have a considerable influence on the production of isobutanol, but the existing research on microbial isobutanol synthesis and product separation [92,93] is weakly linked. This aspect also needs to be strengthened in future research on microbial isobutanol production. We believe that the synthesis of biomass isobutanol will make significant progress with the further research on these issues.

Author Contributions: Conceptualization, Y.S., W.Z. and W.S.; writing—original draft preparation, Y.S. and W.Z.; writing—review and editing, Y.S.; supervision, A.Z. All authors have read and agreed to the published version of the manuscript.

Funding: This work was supported by the Chunhui program of the Ministry of Education (Grant No.Z2017012).

Conflicts of Interest: The authors declare no conflict of interest.

References

1. Cherubini, F.; Strømman, A.H. Life cycle assessment of bioenergy systems: State of the art and future challenges. *Bioresour. Technol.* **2011**, *102*, 437–451. [CrossRef] [PubMed]
2. Lee, S.K.; Chou, H.; Ham, T.S.; Lee, T.S.; Keasling, J.D. Metabolic engineering of microorganisms for biofuels production: From bugs to synthetic biology to fuels. *Curr. Opin. Biotechnol.* **2008**, *19*, 556–563. [CrossRef] [PubMed]
3. Raud, M.; Kikas, T.; Sippula, O.; Shurpali, N. Potentials and challenges in lignocellulosic biofuel production technology. *Renew. Sustain. Energy Rev.* **2019**, *111*, 44–56. [CrossRef]

4. Choi, Y.J.; Lee, J.; Jang, Y.-S.; Lee, S.Y. Metabolic engineering of microorganisms for the production of higher alcohols. *mBio* **2014**, *5*, e01524-14. [CrossRef] [PubMed]

5. Liao, J.C.; Mi, L.; Pontrelli, S.; Luo, J.C.L.L.M.S.P.S. Fuelling the future: Microbial engineering for the production of sustainable biofuels. *Nat. Rev. Genet.* **2016**, *14*, 288–304. [CrossRef]

6. Bilal, M.; Iqbal, H.M.; Hu, H.; Wang, W.; Zhang, X. Metabolic engineering and enzyme-mediated processing: A biotechnological venture towards biofuel production—A review. *Renew. Sustain. Energy Rev.* **2018**, *82*, 436–447. [CrossRef]

7. Isikgor, F.H.; Becer, C.R. Lignocellulosic biomass: A sustainable platform for the production of bio-based chemicals and polymers. *Polym. Chem.* **2015**, *6*, 4497–4559. [CrossRef]

8. Valdivia, M.; Galan, J.L.; Laffarga, J.; Ramos, J. Biofuels 2020: Biorefineries based on lignocellulosic materials. *Microb. Biotechnol.* **2016**, *9*, 585–594. [CrossRef]

9. Lara-Flores, A.A.; Araújo, R.G.; Rodríguez-Jasso, R.M.; Aguedo, M.; Aguilar, C.N.; Trajano, H.L.; Ruiz, H.A. *Bioeconomy and Biorefinery: Valorization of Hemicellulose from Lignocellulosic Biomass and Potential Use of Avocado Residues as a Promising Resource of Bioproducts*; Singhania, R., Agarwal, R., Kumar, R., Sukumaran, R., Eds.; Waste to Wealth; Springer: Singapore, 2018; pp. 141–170.

10. Weber, C.; Farwick, A.; Benisch, F.; Brat, D.; Dietz, H.; Subtil, T.; Boles, E. Trends and challenges in the microbial production of lignocellulosic bioalcohol fuels. *Appl. Microbiol. Biotechnol.* **2010**, *87*, 1303–1315. [CrossRef]

11. Amoah, J.; Kahar, P.; Ogino, C.; Kondo, A. Bioenergy and biorefinery: Feedstock, biotechnological conversion, and products. *Biotechnol. J.* **2019**, *14*, e1800494. [CrossRef]

12. Toor, M.; Kumar, S.S.; Malyan, S.K.; Bishnoi, N.R.; Mathimani, T.; Rajendran, K.; Pugazhendhi, A. An overview on bioethanol production from lignocellulosic feedstocks. *Chemosphere* **2020**, *242*, 125080. [CrossRef] [PubMed]

13. Wang, B.-W.; Shi, A.-Q.; Tu, R.; Zhang, X.-L.; Wang, Q.; Bai, F. Branched-chain higher alcohols. *Process Integr. Biochem. Eng.* **2011**, *128*, 101–118. [CrossRef]

14. Atsumi, S.; Hanai, T.; Liao, J.C. Non-fermentative pathways for synthesis of branched-chain higher alcohols as biofuels. *Nat. Cell Biol.* **2008**, *451*, 86–89. [CrossRef] [PubMed]

15. Connor, M.R.; Liao, J.C. Microbial production of advanced transportation fuels in non-natural hosts. *Curr. Opin. Biotechnol.* **2009**, *20*, 307–315. [CrossRef] [PubMed]

16. Shahsavan, M.; Mack, J.H. Numerical study of a boosted HCCI engine fueled with n-butanol and isobutanol. *Energy Convers. Manag.* **2018**, *157*, 28–40. [CrossRef]

17. Savage, N. Fuel options: The ideal biofuel. *Nat. Cell Biol.* **2011**, *474*, S9–S11. [CrossRef]

18. Elfasakhany, A. Investigations on performance and pollutant emissions of spark-ignition engines fueled with n-butanol-, isobutanol-, ethanol-, methanol-, and acetone-gasoline blends: A comparative study. *Renew. Sustain. Energy Rev.* **2017**, *71*, 404–413. [CrossRef]

19. Baez, A.; Cho, K.-M.; Liao, J.C. High-flux isobutanol production using engineered *Escherichia coli*: A bioreactor study with in situ product removal. *Appl. Microbiol. Biotechnol.* **2011**, *90*, 1681–1690. [CrossRef]

20. Smith, K.M.; Liao, J.C. An evolutionary strategy for isobutanol production strain development in *Escherichia coli*. *Metab. Eng.* **2011**, *13*, 674–681. [CrossRef]

21. Bastian, S.; Liu, X.; Meyerowitz, J.T.; Snow, C.D.; Chen, M.M.; Arnold, F.H. Engineered ketol-acid reductoisomerase and alcohol dehydrogenase enable anaerobic 2-methylpropan-1-ol production at theoretical yield in *Escherichia coli*. *Metab. Eng.* **2011**, *13*, 345–352. [CrossRef]

22. Smith, K.M.; Cho, K.-M.; Liao, J.C. Engineering *Corynebacterium glutamicum* for isobutanol production. *Appl. Microbiol. Biotechnol.* **2010**, *87*, 1045–1055. [CrossRef] [PubMed]

23. Blombach, B.; Riester, T.; Wieschalka, S.; Ziert, C.; Youn, J.-W.; Wendisch, V.F.; Eikmanns, B.J. *Corynebacterium glutamicum* tailored for efficient isobutanol production. *Appl. Environ. Microbiol.* **2011**, *77*, 3300–3310. [CrossRef] [PubMed]

24. Hasegawa, S.; Jojima, T.; Suda, M.; Inui, M. Isobutanol production in *Corynebacterium glutamicum*: Suppressed succinate by-production by pckA inactivation and enhanced productivity via the Entner-Doudoroff pathway. *Metab. Eng.* **2020**, *59*, 24–35. [CrossRef] [PubMed]

25. Li, S.; Wen, J.; Jia, X. Engineering *Bacillus subtilis* for isobutanol production by heterologous Ehrlich pathway construction and the biosynthetic 2-ketoisovalerate precursor pathway overexpression. *Appl. Microbiol. Biotechnol.* **2011**, *91*, 577–589. [CrossRef] [PubMed]

26. Shouliang, Y.; Huang, D.; Li, Y.; Wen, J.; Jia, X. Rational improvement of the engineered isobutanol-producing *Bacillus subtilis* by elementary mode analysis. *Microb. Cell Factories* **2012**, *11*, 101. [CrossRef]

27. Qi, H.; Li, S.; Zhao, S.; Huang, D.; Xia, M.; Wen, J. Model-driven redox pathway manipulation for improved isobutanol production in *bacillus subtilis* complemented with experimental validation and metabolic profiling analysis. *PLoS ONE* **2014**, *9*, e93815. [CrossRef]

28. Liu, S.; Qureshi, N. How microbes tolerate ethanol and butanol. *New Biotechnol.* **2009**, *26*, 117–121. [CrossRef]

29. Chen, X.; Nielsen, K.F.; Borodina, I.; Kielland-Brandt, M.C.; Karhumaa, K. Increased isobutanol production in *Saccharomyces cerevisiae* by overexpression of genes in valine metabolism. *Biotechnol. Biofuels* **2011**, *4*, 21. [CrossRef]

30. Brat, D.; Weber, C.; Lorenzen, W.; Bode, H.B.; Boles, E. Cytosolic re-localization and optimization of valine synthesis and catabolism enables inseased isobutanol production with the yeast *Saccharomyces cerevisiae*. *Biotechnol. Biofuels* **2012**, *5*, 65. [CrossRef]

31. Wess, J.; Brinek, M.; Boles, E. Improving isobutanol production with the yeast *Saccharomyces cerevisiae* by successively blocking competing metabolic pathways as well as ethanol and glycerol formation. *Biotechnol. Biofuels* **2019**, *12*, 1–15. [CrossRef]

32. Zhao, E.M.; Zhang, Y.; Mehl, J.; Park, H.; Lalwani, M.A.; Toettcher, J.E.; Avalos, J.L. Optogenetic regulation of engineered cellular metabolism for microbial chemical production. *Nat. Cell Biol.* **2018**, *555*, 683–687. [CrossRef] [PubMed]

33. Ferreira, J.A.; Brancoli, P.; Agnihotri, S.; Bolton, K.; Taherzadeh, M.J. A review of integration strategies of lignocelluloses and other wastes in 1st generation bioethanol processes. *Process. Biochem.* **2018**, *75*, 173–186. [CrossRef]

34. Liu, C.-G.; Xiao, Y.; Xia, X.-X.; Zhao, X.-Q.; Peng, L.; Srinophakun, P.; Bai, F. Cellulosic ethanol production: Progress, challenges and strategies for solutions. *Biotechnol. Adv.* **2019**, *37*, 491–504. [CrossRef]

35. Xin, F.; Dong, W.; Zhang, W.; Ma, J.; Jiang, M. Biobutanol production from crystalline cellulose through consolidated bioprocessing. *Trends Biotechnol.* **2019**, *37*, 167–180. [CrossRef] [PubMed]

36. Wen, Z.; Li, Q.; Liu, J.; Jin, M.; Yang, S. Consolidated bioprocessing for butanol production of cellulolytic Clostridia: Development and optimization. *Microb. Biotechnol.* **2019**, *13*, 410–422. [CrossRef] [PubMed]

37. Tao, L.; Tan, E.C.D.; McCormick, R.L.; Zhang, M.; Aden, A.; He, X.; Zigler, B.T. Techno-economic analysis and life-cycle assessment of cellulosic isobutanol and comparison with cellulosic ethanol and n-butanol. *Biofuels Bioprod. Biorefining* **2014**, *8*, 30–48. [CrossRef]

38. Hal, W.J. *Iso-Butanol Platform Rotterdam (IBPR)*; Policy Studies; ECN Biomass & Energy Efficiency: Schagen, The Netherlands, 2016.

39. Higashide, W.; Li, Y.; Yang, Y.; Liao, J.C. Metabolic engineering of *Clostridium cellulolyticum* for production of isobutanol from cellulose. *Appl. Environ. Microbiol.* **2011**, *77*, 2727–2733. [CrossRef]

40. Guedon, E.; Desvaux, M.; Payot, S.; Petitdemange, H. Growth inhibition of *Clostridium cellulolyticum* by an inefficiently regulated carbon flow. *Microbiology* **1999**, *145*, 1831–1838. [CrossRef]

41. Li, Y.; Xu, T.; Tschaplinski, T.J.; Engle, N.L.; Yang, Y.; Graham, D.E.; He, Z.; Zhou, J. Improvement of cellulose catabolism in *Clostridium cellulolyticum* by sporulation abolishment and carbon alleviation. *Biotechnol. Biofuels* **2014**, *7*, 25. [CrossRef]

42. Lin, P.P.; Rabe, K.S.; Takasumi, J.L.; Kadisch, M.; Arnold, F.H.; Liao, J.C. Isobutanol production at elevated temperatures in thermophilic *Geobacillus thermoglucosidasius*. *Metab. Eng.* **2014**, *24*, 1–8. [CrossRef]

43. Lin, P.P.; Mi, L.; Morioka, A.H.; Yoshino, K.M.; Konishi, S.; Xu, S.C.; Papanek, B.A.; Riley, L.A.; Guss, A.M.; Liao, J.C. Consolidated bioprocessing of cellulose to isobutanol using *Clostridium thermocellum*. *Metab. Eng.* **2015**, *31*, 44–52. [CrossRef]

44. Holwerda, E.K.; Olson, D.G.; Ruppertsberger, N.M.; Stevenson, D.M.; Murphy, S.J.L.; Maloney, M.I.; Lanahan, A.A.; Amador-Noguez, D.; Lynd, L.R. Metabolic and evolutionary responses of *Clostridium thermocellum* to genetic interventions aimed at improving ethanol production. *Biotechnol. Biofuels* **2020**, *13*, 1–20. [CrossRef]

45. Minty, J.J.; Singer, M.E.; Scholz, S.A.; Bae, C.-H.; Ahn, J.-H.; Foster, C.E.; Liao, J.C.; Lin, X.N. Design and characterization of synthetic fungal-bacterial consortia for direct production of isobutanol from cellulosic biomass. *Proc. Nat. Acad. Sci. USA* **2013**, *110*, 14592–14597. [CrossRef]

46. Rubinstein, G.M.; Lipscomb, G.L.; Williams-Rhaesa, A.M.; Schut, G.J.; Kelly, R.M.; Adams, M.W.W. Engineering the cellulolytic extreme thermophile *Caldicellulosiruptor bescii* to reduce carboxylic acids to alcohols using plant biomass as the energy source. *J. Ind. Microbiol. Biotechnol.* **2020**, *47*, 1–13. [CrossRef]

47. Nakashima, N.; Tamura, T. A new carbon catabolite repression mutation of *Escherichia coli*, mlc, and its use for producing isobutanol. *J. Biosci. Bioeng.* **2012**, *114*, 38–44. [CrossRef] [PubMed]

48. Akita, H.; Nakashima, N.; Hoshino, T. Bacterial production of isobutanol without expensive reagents. *Appl. Microbiol. Biotechnol.* **2014**, *99*, 991–999. [CrossRef] [PubMed]

49. Brat, D.; Boles, E. Isobutanol production fromd-xylose by recombinant *Saccharomyces cerevisiae*. *FEMS Yeast Res.* **2013**, *13*, 241–244. [CrossRef] [PubMed]

50. Zhou, H.; Cheng, J.-S.; Wang, B.L.; Fink, G.R.; Stephanopoulos, G. Xylose isomerase overexpression along with engineering of the pentose phosphate pathway and evolutionary engineering enable rapid xylose utilization and ethanol production by *Saccharomyces cerevisiae*. *Metab. Eng.* **2012**, *14*, 611–622. [CrossRef]

51. Zhang, Y.; Lane, S.; Chen, J.-M.; Hammer, S.K.; Luttinger, J.; Yang, L.; Jin, Y.-S.; Avalos, J.L. Xylose utilization stimulates mitochondrial production of isobutanol and 2-methyl-1-butanol in *Saccharomyces cerevisiae*. *Biotechnol. Biofuels* **2019**, *12*, 1–15. [CrossRef]

52. Promdonkoy, P.; Mhuantong, W.; Champreda, V.; Tanapongpipat, S.; Runguphan, W. Improvement in d-xylose utilization and isobutanol production in *S. cerevisiae* by adaptive laboratory evolution and rational engineering. *J. Ind. Microbiol. Biotechnol.* **2020**, *47*, 497–510. [CrossRef]

53. Kim, S.R.; Skerker, J.M.; Kang, W.; Lesmana, A.; Wei, N.; Arkin, A.P.; Jin, Y.-S. Rational and evolutionary engineering approaches uncover a small set of genetic changes efficient for rapid xylose fermentation in *Saccharomyces cerevisiae*. *PLoS ONE* **2013**, *8*, e57048. [CrossRef] [PubMed]

54. Lane, S.; Zhang, Y.; Yun, E.J.; Ziolkowski, L.; Zhang, G.; Jin, Y.-S.; Avalos, J.L. Xylose assimilation enhances the production of isobutanol in engineered *Saccharomyces cerevisiae*. *Biotechnol. Bioeng.* **2019**, *117*, 372–381. [CrossRef] [PubMed]

55. Lange, J.; Müller, F.; Takors, R.; Blombach, B. Harnessing novel chromosomal integration loci to utilize an organosolv-derived hemicellulose fraction for isobutanol production with engineered *Corynebacterium glutamicum*. *Microb. Biotechnol.* **2017**, *11*, 257–263. [CrossRef] [PubMed]

56. Krause, F.S.; Blombach, B.; Eikmanns, B.J. Metabolic Engineering of *Corynebacterium glutamicum* for 2-Ketoisovalerate Production. *Appl. Environ. Microbiol.* **2010**, *76*, 8053–8061. [CrossRef]

57. Desai, S.H.; Rabinovitch-Deere, C.A.; Tashiro, Y.; Atsumi, S. Isobutanol production from cellobiose in *Escherichia coli*. *Appl. Microbiol. Biotechnol.* **2014**, *98*, 3727–3736. [CrossRef]

58. Desai, S.H.; A Rabinovitch-Deere, C.; Fan, Z.; Atsumi, S. Isobutanol production from cellobionic acid in *Escherichia coli*. *Microb. Cell Factories* **2015**, *14*, 1–10. [CrossRef]

59. Su, H.; Lin, J.; Wang, G. Metabolic engineering of *Corynebacterium crenatium* for enhancing production of higher alcohols. *Sci. Rep.* **2016**, *6*, 39543. [CrossRef] [PubMed]

60. Yang, J.; Kim, J.K.; Ahn, J.-O.; Song, Y.-H.; Shin, C.-S.; Park, Y.-C.; Kim, K.H. Isobutanol production from empty fruit bunches. *Renew. Energy* **2020**, *157*, 1124–1130. [CrossRef]

61. Jung, H.-M.; Lee, J.Y.; Lee, J.-H.; Oh, M.-K. Improved production of isobutanol in pervaporation-coupled bioreactor using sugarcane bagasse hydrolysate in engineered *Enterobacter aerogenes*. *Bioresour. Technol.* **2018**, *259*, 373–380. [CrossRef]

62. Jung, H.M.; Kim, Y.H.; Oh, M.K. Formate and nitrate utilization in *Enterobacter aerogenes* for semi-anaerobic production of isobutanol. *Biotechnol. J.* **2017**, *12*, 1700121. [CrossRef]

63. Huo, Y.-X.; Cho, K.M.; Rivera, J.G.L.; Monte, E.; Shen, C.R.; Yan, Y.; Liao, J.C. Conversion of proteins into biofuels by engineering nitrogen flux. *Nat. Biotechnol.* **2011**, *29*, 346–351. [CrossRef]

64. Wu, W.; Tran-Gyamfi, M.B.; Jaryenneh, J.D.; Davis, R.W. Cofactor engineering of ketol-acid reductoisomerase (IlvC) and alcohol dehydrogenase (YqhD) improves the fusel alcohol yield in algal protein anaerobic fermentation. *Algal Res.* **2016**, *19*, 162–167. [CrossRef]

65. Liu, F.; Wu, W.; Tran-Gymfi, M.B.; Jaryenneh, J.D.; Zhuang, X.; Davis, R.W. Bioconversion of distillers' grains hydrolysates to advanced biofuels by an *Escherichia coli* co-culture. *Microb. Cell Factor.* **2017**, *16*, 192. [CrossRef] [PubMed]

66. Kim, E.-J.; Seo, D.; Choi, K.-Y. Bioalcohol production from spent coffee grounds and okara waste biomass by engineered *Bacillus subtilis*. *Biomass. Convers. Biorefinery* **2019**, *10*, 167–173. [CrossRef]

67. Tian, L.; Papanek, B.; Olson, D.G.; Rydzak, T.; Holwerda, E.K.; Zheng, T.; Zhou, J.; Maloney, M.; Jiang, N.; Giannone, R.J.; et al. Simultaneous achievement of high ethanol yield and titer in *Clostridium thermocellum*. *Biotechnol. Biofuels* **2016**, *9*, 1–11. [CrossRef] [PubMed]

68. Tripathi, S.A.; Olson, D.G.; Argyros, D.A.; Miller, B.B.; Barrett, T.F.; Murphy, D.M.; McCool, J.D.; Warner, A.K.; Rajgarhia, V.B.; Lynd, L.R.; et al. Development of pyrF-based genetic system for targeted gene deletion in *Clostridium thermocellum* and creation of a pta mutant. *Appl. Environ. Microbiol.* **2010**, *76*, 6591–6599. [CrossRef] [PubMed]

69. Olson, D.G.; Lynd, L.R. Transformation of *Clostridium thermocellum* by electroporation. *Cellulases* **2012**, *510*, 317–330. [CrossRef]

70. Mearls, E.B.; Olson, D.G.; Herring, C.D.; Lynd, L.R. Development of a regulatable plasmid-based gene expression system for *Clostridium thermocellum*. *Appl. Microbiol. Biotechnol.* **2015**, *99*, 7589–7599. [CrossRef]

71. Walker, J.E.; Lanahan, A.A.; Zheng, T.; Toruno, C.; Lynd, L.R.; Cameron, J.C.; Olson, D.G.; Eckert, C.A. Development of both type I-B and type II CRISPR/Cas genome editing systems in the cellulolytic bacterium *Clostridium thermocellum*. *Metab. Eng. Commun.* **2020**, *10*, e00116. [CrossRef]

72. Zhao, C.; Zhang, Y.; Li, Y. Production of fuels and chemicals from renewable resources using engineered *Escherichia coli*. *Biotechnol. Adv.* **2019**, *37*, 107402. [CrossRef]

73. Hasunuma, T.; Ishii, J.; Kondo, A. Rational design and evolutional fine tuning of *Saccharomyces cerevisiae* for biomass breakdown. *Curr. Opin. Chem. Biol.* **2015**, *29*, 1–9. [CrossRef] [PubMed]

74. Sharma, S.; Arora, A. Tracking strategic developments for conferring xylose utilization/fermentation by *Saccharomyces cerevisiae*. *Ann. Microbiol.* **2020**, *70*, 1–17. [CrossRef]

75. Promdonkoy, P.; Siripong, W.; Downes, J.J.; Tanapongpipat, S.; Runguphan, W. Systematic improvement of isobutanol production from d-xylose in engineered *Saccharomyces cerevisiae*. *AMB Express* **2019**, *9*, 1–14. [CrossRef]

76. Bengtsson, O.; Hahn-Hägerdal, B.; Gorwa-Grauslund, M.F. Xylose reductase from Pichia stipitis with altered coenzyme preference improves ethanolic xylose fermentation by recombinant *Saccharomyces cerevisiae*. *Biotechnol. Biofuels* **2009**, *2*, 9. [CrossRef]

77. Petschacher, B.; Nidetzky, B. Altering the coenzyme preference of xylose reductase to favor utilization of NADH enhances ethanol yield from xylose in a metabolically engineered strain of *Saccharomyces cerevisiae*. *Microb. Cell Factories* **2008**, *7*, 9. [CrossRef] [PubMed]

78. Watanabe, S.; Abu Saleh, A.; Pack, S.P.; Annaluru, N.; Kodaki, T.; Makino, K. Ethanol production from xylose by recombinant *Saccharomyces cerevisiae* expressing protein engineered NADP+ -dependent xylitol dehydrogenase. *J. Biotechnol.* **2007**, *130*, 316–319. [CrossRef] [PubMed]

79. Chen, Y.; Wu, Y.; Zhu, B.; Zhang, G.; Wei, N. Co-fermentation of cellobiose and xylose by mixed culture of recombinant *Saccharomyces cerevisiae* and kinetic modeling. *PLoS ONE* **2018**, *13*, e0199104. [CrossRef]

80. Vinuselvi, P.; Lee, S.K. Engineered *Escherichia coli* capable of co-utilization of cellobiose and xylose. *Enzym. Microb. Technol.* **2012**, *50*, 1–4. [CrossRef]

81. Kumar, B.; Bhardwaj, N.; Agrawal, K.; Chaturvedi, V.; Verma, P. Current perspective on pretreatment technologies using lignocellulosic biomass: An emerging biorefinery concept. *Fuel Process. Technol.* **2020**, *199*, 106244. [CrossRef]

82. Jönsson, L.J.; Alriksson, B.; Nilvebrant, N.-O. Bioconversion of lignocellulose: Inhibitors and detoxification. *Biotechnol. Biofuels* **2013**, *6*, 16. [CrossRef] [PubMed]

83. Su, H.; Jiang, J.; Lu, Q.; Zhao, Z.; Xie, T.; Zhao, H.; Wang, M. Engineering *Corynebacterium crenatum* to produce higher alcohols for biofuel using hydrolysates of duckweed (Landoltia punctata) as feedstock. *Microb. Cell Factories* **2015**, *14*, 16. [CrossRef] [PubMed]

84. Choi, K.-Y.; Wernick, D.G.; Tat, C.A.; Liao, J.C. Consolidated conversion of protein waste into biofuels and ammonia using *Bacillus subtilis*. *Metab. Eng.* **2014**, *23*, 53–61. [CrossRef]

85. Tojo, S.; Satomura, T.; Kumamoto, K.; Hirooka, K.; Fujita, Y. Molecular mechanisms underlying the positive stringent response of the *Bacillus subtilis* ilv-leu operon, involved in the biosynthesis of branched-chain amino acids. *J. Bacteriol.* **2008**, *190*, 6134–6147. [CrossRef] [PubMed]

86. Kim, S.Y.; Yang, Y.-H.; Choi, K.-Y. Bioconversion of plant hydrolysate biomass into biofuels using an engineered *Bacillus subtilis* and *Escherichia coli* mixed-whole cell biotransformation. *Biotechnol. Bioprocess. Eng.* **2020**, *25*, 477–484. [CrossRef]

87. Robak, K.; Balcerek, M. Current state-of-the-art in ethanol production from lignocellulosic feedstocks. *Microbiol. Res.* **2020**, *240*, 126534. [CrossRef]

88. Kuroda, K.; Hammer, S.K.; Watanabe, Y.; López, J.M.; Fink, G.R.; Stephanopoulos, G.; Ueda, M.; Avalos, J.L. Critical roles of the pentose phosphate pathway and GLN3 in isobutanol-specific tolerance in yeast. *Cell Syst.* **2019**, *9*, 534–547.e5. [CrossRef]

89. Seo, H.-M.; Jeon, J.-M.; Lee, J.H.; Song, H.-S.; Joo, H.-B.; Park, S.-H.; Choi, K.-Y.; Kim, Y.H.; Park, K.; Ahn, J.; et al. Combinatorial application of two aldehyde oxidoreductases on isobutanol production in the presence of furfural. *J. Ind. Microbiol. Biotechnol.* **2015**, *43*, 37–44. [CrossRef]

90. Song, H.-S.; Jeon, J.-M.; Choi, Y.-K.; Kim, J.-Y.; Kim, W.; Rene, E.R.; Park, K.; Ahn, J.-O.; Lee, H.; Yang, Y.-H. L-Glycine alleviates furfural-induced growth inhibition during isobutanol production in *Escherichia coli*. *J. Microbiol. Biotechnol.* **2017**, *27*, 2165–2172. [CrossRef]

91. Song, H.-S.; Jeon, J.-M.; Kim, H.-J.; Bhatia, S.; Sathiyanarayanan, G.; Kim, J.; Hong, J.W.; Hong, Y.G.; Choi, K.-Y.; Kim, Y.-G.; et al. Increase in furfural tolerance by combinatorial overexpression of NAD salvage pathway enzymes in engineered isobutanol-producing *E. coli*. *Bioresour. Technol.* **2017**, *245*, 1430–1435. [CrossRef]

92. Andre, A.; Nagy, T.; Toth, A.J.; Haaz, E.; Fozer, D.; Tarjani, J.A.; Mizsey, P. Distillation contra pervaporation: Comprehensive investigation of isobutanol-water separation. *J. Clean. Prod.* **2018**, *187*, 804–818. [CrossRef]

93. Farhadi, M.; Pazuki, G.; Raisi, A. Modeling of the pervaporation process for isobutanol purification from aqueous solution using intelligent systems. *Sep. Sci. Technol.* **2017**, *53*, 1383–1396. [CrossRef]

Publisher's Note: MDPI stays neutral with regard to jurisdictional claims in published maps and institutional affiliations.

*applied
sciences*

Article

Observation of Potential Contaminants in Processed Biomass Using Fourier Transform Infrared Spectroscopy

Jingshun Zhuang [1], Mi Li [2], Yunqiao Pu [3], Arthur Jonas Ragauskas [2,3,4,*] and Chang Geun Yoo [1,3,*]

1   Department of Chemical Engineering, State University of New York College of Environmental Science and Forestry, Syracuse, NY 13104, USA; jzhuan02@syr.edu
2   Center for Renewable Carbon, Department of Forestry, Wildlife, and Fisheries, University of Tennessee Institute of Agriculture, Knoxville, TN 37996, USA; mli47@utk.edu
3   Center for Bioenergy Innovation, Biosciences Division, University of Tennessee-Oak Ridge National Laboratory Joint Institute for Biological Science, Oak Ridge National Laboratory, Oak Ridge, TN 37831, USA; puy1@ornl.gov
4   Department of Chemical and Biomolecular Engineering, University of Tennessee, Knoxville, TN 37996, USA
*   Correspondence: aragausk@utk.edu (A.J.R.); cyoo05@esf.edu (C.G.Y.)

Received: 2 June 2020; Accepted: 22 June 2020; Published: 24 June 2020

Abstract: With rapidly increased interests in biomass, diverse chemical and biological processes have been applied for biomass utilization. Fourier transform infrared (FTIR) analysis has been used for characterizing different types of biomass and their products, including natural and processed biomass. During biomass treatments, some solvents and/or catalysts can be retained and contaminate biomass. In addition, contaminants can be generated by the decomposition of biomass components. Herein, we report FTIR analyses of a series of contaminants, such as various solvents, chemicals, enzymes, and possibly formed degradation by-products in the biomass conversion process along with poplar biomass. This information helps to prevent misunderstanding the FTIR analysis results of the processed biomass.

Keywords: poplar; FTIR; contaminants; by-products

---

1. Introduction

A proper understanding of biomass characteristics is important for the effective utilization of biomass. It not only provides natural properties of biomass but also tells the influences of the applied processes on the biomass structures. Characteristics of biomass have been investigated in different aspects, including physical, chemical, thermal, mineral, and surface properties. For a better understanding of biomass and its products, several different analytical approaches have been developed using diverse analytical techniques such as high-performance liquid chromatography (HPLC), gas chromatography (GC), gel permeation chromatography (GPC), nuclear magnetic resonance (NMR), time-of-flight secondary ion mass spectrometry (ToF-SIMS), X-ray, transmission electron microscopy (TEM), scanning electron microscopy (SEM), differential scanning calorimetry (DSC), thermogravimetric analysis (TGA), and Fourier transform infrared (FTIR) for measuring carbohydrate contents, identification and quantity of products, molecular weights, structural information such as linkages and composition, spatial distribution of molecules and chemical structures on surface, crystallinity, morphological characteristics in nano- and micro-scales, thermal properties (melting point, glass transition temperature, etc.), functional groups and chemical bonds, and other important information of biomass and its products and by-products [1–9]. Among these methods,

FTIR spectroscopy is one of the most widely applied analytical methods to study the functional groups of biomass by measuring the absorption bands of samples [10]. It provides qualitative and semi-quantitative information for functional groups of biomass by determining the presence of fundamental molecular vibrations [11]. Its Fellgett and Jacquinot advantages allow for rapid and ready characterization compared to many other biomass analysis methods [12]. Moreover, it does not need any modification and/or deconstruction of biomass; therefore, original properties can be monitored as the sample is. Despite these advantages, the characterization of biomass using FTIR is still challenged by overlapping the bands from different biomass components and/or unexpected impurities from the applied catalysts and solvents. In particular, fingerprint regions are complicated to identify because of many series of absorptions. For fast and reliable analysis of the substances from different processing, detection, and identification of possible contaminants are very important.

Lignocellulosic biomass is a heterogeneous matrix. Due to the complicated composition and structural properties of biomass, single or multi-stage pretreatment/preprocessing is necessary for its utilization, isolation, and analyses. Various chemicals, such as organic solvents, acids, alkalis, and inorganic salt solutions have been applied for isolation, pretreatment, conversion, and other reactions on biomass [13–19]. Biological catalysts, such as enzymes, have also been used in many biomass conversion processes or characterization methods [20]. Besides, each biomass component could be decomposed and/or modified under severe process conditions [21]. The presence of these chemicals and by-products are considered as impurities and could potentially affect their characterization results; therefore, they should be completely removed after the processes. Unfortunately, these components are possibly retained on the surface of biomass after these preprocessing and cause misinterpretation of the targeted biomass structure by their overlapped FTIR spectra. Besides the misreading of the biomass properties, the detection of contaminants can be used to determine the necessity of biomass washing step. Although the IR assignments of many chemicals and solvents are available individually, their actual contaminations are not easily detected due to the spectra of biomass itself. In this study, poplar biomass was mixed with known chemicals and enzymes, which are potential contaminants, and their overlapped FTIR spectra in each sample were identified and discussed.

2. Materials and Methods

2.1. Materials

Poplar was harvested in the Oak Ridge National Laboratory in 2008. Prior to the FTIR analysis, the sample was Wiley-milled and screened to 0.42 mm. Extractives were removed from the original poplar sample (~10 g) by toluene/ethanol Soxhlet extraction (2:1, *v/v*, 200 mL) for 8 h followed by 6 h of water extraction. All chemicals (acetone, ethanol, methanol, tetrahydrofuran, dioxane, toluene, glycerol, chloroform, pyridine, sulfuric acid, hydrochloric acid, phosphoric acid, acetic acid, sodium hydroxide, ammonium hydroxide, 1-butyl-3-methylimidazolium chloride, 1-benzyl-3-methylimidazolium chloride, choline chloride, urea, *p*-hydroxybenzoic acid, 4-hydroxybenzaldehyde, *p*-coumaric acid, hydroxymethylfurfural, and furfural) and enzymes (cellulase and β-glucosidase) used in this study were purchased from VWR, Sigma-Aldrich, or Fisher Scientific. Deep eutectic solvents (DESs) formed by combining hydrogen bonding donors (HBDs: urea, *p*-hydroxybenzoic acid, 4-hydroxybenzaldehyde, *p*-coumaric acid) and hydrogen bonding acceptor (HBA: choline chloride) at 80 °C prior to the FTIR analysis.

2.2. Isolation of Cellulose, Hemicellulose, and Lignin

Cellulose, hemicellulose, and lignin were isolated from the extractives-free poplar, as described in the previous studies [2,22]. In brief, the biomass was delignified using peracetic acid at 25 °C with 5% (wt/wt) solid loading for 24 h. The remaining solid, holocellulose, was air dried for 24 h. Two-step alkali extraction with 17.5% (wt/wt) and 8.75% (wt/wt) sodium hydroxide was conducted at 25 °C for 2 h in each step. The remaining solid fraction was called α-cellulose after being air dried, and the liquid fraction was neutralized with anhydrous acetic acid and mixed with ethanol three times to precipitate hemicellulose.

Cellulolytic enzyme lignin (CEL) was separated from the poplar samples. The extractives-free poplar was ball-milled using Retsch PM 100 at 600 rpm for 2 h. The ball-milled sample was hydrolyzed at 50 °C with the CTec2 enzyme in acetate buffer solution (pH 4.8) for 48 h twice. The residual solid was extracted with 96% dioxane for 48 h. The dioxane-extracted fraction was recovered at 40 °C by rotary evaporation and freeze drying and used for further analysis.

2.3. Fourier Transform Infrared (FTIR) Analysis

To observe the FTIR spectra of contaminants from the spectra of biomass clearly, about 30–50 µL of contaminant was loaded to 0.3 g of extractives-free poplar in 20 mL glass vial and mixed by vortexing prior to the analysis. The prepared DESs were loaded to biomass and physically mixed using a glass rod due to their relatively high viscosity. FTIR analysis was conducted using the Spectrum One FTIR spectrometer (PerkinElmer, Wellesley, MA, USA) equipped with a universal attenuated total reflection (ATR) accessory. ATR-FTIR spectra between 4000 and 600 cm^{-1} were measured at a 4 cm^{-1} resolution and averaging 16 scans per sample.

3. Results and Discussion

3.1. Major Components of Biomass

Cellulose, hemicellulose, and lignin are three main components in lignocellulosic biomass. Table 1 and Figure 1 present the FTIR band assignments and spectra of poplar and its major components isolated from the same biomass. Prior to the analysis, other extractives in the poplar sample were removed by two-step extraction: 8 h toluene/ethanol Soxhlet-extraction followed by 6 h water extraction. Isolated cellulose, hemicellulose, and lignin were analyzed using FTIR and compared with the extractives-free poplar. The assignment of each band was identified according to the previous studies [23–31].

Table 1. Fourier transform infrared (FTIR) band assignments of poplar and its major components: cellulose, hemicellulose, and lignin [23–31].

	Wavenumber [cm^{-1}]	Assignment	Components
1	3367	O-H stretching	Cellulose, Hemicellulose, Lignin
2	2914	C-H stretching	Cellulose, Hemicellulose, Lignin
3	1745	C=O stretching	Hemicellulose, Lignin
4	1618	Aromatic skeletal vibration, C=O stretching, adsorbed O-H	Hemicellulose, Lignin
5	1508	C=C-C aromatic ring stretching and vibration	Lignin
6	1457	C-H deformation (in methyl and methylene)	Lignin
7	1424	Symmetric CH$_2$ bending vibration, symmetric stretching band of carboxyl group, C-H deformation	Cellulose, Hemicellulose, Lignin
8	1370	C-H bending, C-H stretching in CH$_3$	Cellulose, Hemicellulose, Lignin
9	1317	CH$_2$ wagging, C-O stretching of C$_5$ substituted aromatic units	Cellulose, Hemicellulose, Lignin
10	1235	C-O stretching of guaiacyl unit	Lignin
11	1160	C-O-C stretching	Cellulose, Hemicellulose
12	1108	Aromatic C-H in plane deformation	Lignin
13	1053	C-OH stretching vibration, C-O deformation	Cellulose, Hemicellulose, Lignin
14	1032	C-O stretching, aromatic C-H in plane deformation	Cellulose, Lignin
15	896	C-O-C stretching	Cellulose, hemicellulose
16	846	Aromatic C-H out of plane bending	Lignin

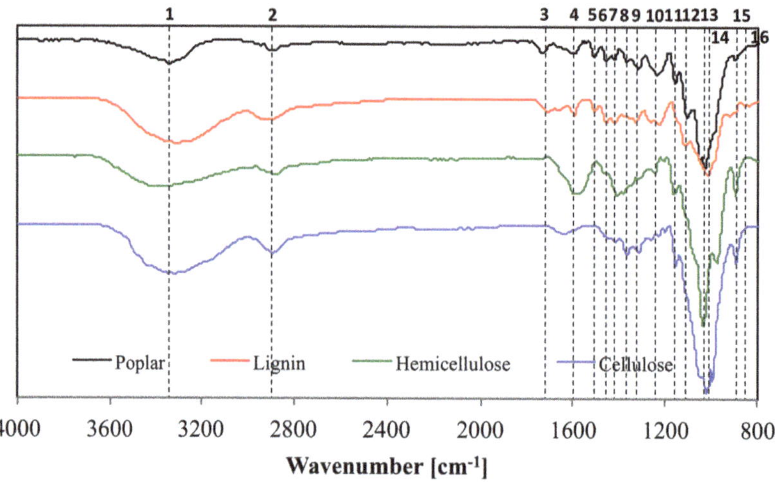

Figure 1. FTIR spectra of poplar and its major components: cellulose, hemicellulose, and lignin. (Note: The assignments of the numbered bands in the figure are described in Table 1).

The IR spectra of poplar and its components showed strong O-H stretching and C-H stretching absorptions at 3367 and 2914 cm^{-1}, respectively. These two strong absorptions are because all three major components in biomass (cellulose, hemicellulose, and lignin) have hydroxy groups and many C-H bonds in their structures. The absorption at 1745 cm^{-1} was due to C=O stretching in hemicellulose and lignin. The absorption at 1618 cm^{-1} represented asymmetric stretching band of the carboxyl group of glucuronic acid in hemicellulose and C=O stretching in conjugated carbonyl of lignin. The band at 1650 cm^{-1} in the IR spectrum of cellulose was possibly caused by adsorbed H_2O. Higher absorption at 3367 cm^{-1} was also observed because of the moisture content in the biomass. In addition, the bands due to symmetric CH_2 bending vibration in cellulose, carboxyl vibration in glucuronic acid with xylan, and C-H in plane deformation with aromatic ring stretching in lignin were observed at 1424 cm^{-1}. The IR absorption bands at 1582 and 1508 cm^{-1} assigned to aromatic ring stretching and vibration (C=C-C) in lignin. The band at 1457 cm^{-1} was observed in lignin due to its C-H deformation in methyl and methylene. The C-H bending in cellulose, hemicellulose, and lignin (aliphatic C-H stretching in methyl and phenolic alcohol) was observed at 1370 cm^{-1}. The CH_2 wagging in cellulose and hemicellulose and the C-O stretching of C_5 substituted aromatic units, such as syringyl and condensed guaiacyl units, were assigned at 1317 cm^{-1}. Similarly, the C-O stretching of guaiacyl unit in lignin was assigned at 1235 cm^{-1}. The bands at 1160 and 896 cm^{-1} arise from C-O-C stretching at the β-(1→4)-glycosidic linkages in cellulose and hemicellulose. The absorption at 1108 cm^{-1} was associated with aromatic C-H in plane deformation for the syringyl unit. The band at 1053 cm^{-1} was assigned to the C-OH stretching vibration of cellulose and hemicellulose. Moreover, this band was for C-O deformation in secondary alcohols and aliphatic ethers. The C-O stretching of cellulose and primary alcohols and C-H in plane deformation for guaiacyl unit exhibited at 1032 cm^{-1}. Aromatic C-H out of plane bending in lignin was presented at 846 cm^{-1}. Although several FTIR bands of different biomass components were overlapped, the IR spectra of samples still provide important clues, including changes of chemical composition, functionalization, and other transformation of biomass for understanding the applied biomass processing.

3.2. Commonly Used Pretreatment and Preprocessing Solvents

Table 2 and Figure S1 show the band assignment for common biomass processing solvents. Water is the most common solution in biomass analysis and the conversion processes. It also exists in the air, and a certain amount can be accumulated in biomass during its storing and processing. The existence of water in biomass remarkably increased the bands at 3354 and 1653 cm^{-1} because of its O-H stretching and O-H-O scissors bending, respectively [32]. Acetone, ethanol, and methanol are common organic solvents for the diverse pre- and post-processing of biomass, and they are also produced from biomass [33]. Acetone contamination on poplar was observed at 3005, 2908, 1713, 1431, 1364, and 1222 cm^{-1} representing its CH_3 degenerated stretching, CH_3 symmetrical stretching, C=O stretching, CH_3 degenerated deformation, CH_3 symmetrical deforming, and C-C stretching, respectively [34]. A decrease in the bands at 3354 and 1653 cm^{-1} is possibly due to the displacement of water in biomass by acetone. The spectra of ethanol impurity were shown at 3350, 2980, and 1056 cm^{-1} for O-H stretching, C-H stretching, and C-O stretching, and those of methanol were at 3352, 2952, 2879, 1465, 1450, 1336, 1053 and 1026 cm^{-1} for O-H stretching, C-H stretching (asymmetric), C-H stretching (symmetric), C-H bending (asymmetric), C-H bending (symmetric), O-H bending, CH_3 rocking, and C-O stretching, respectively [34–36]. Besides these chemicals, tetrahydrofuran (THF), dioxane, toluene, glycerol, pyridine, and chloroform are well-known solvents for diverse biomass pretreatment, isolation/purification, and analyses [1,3,19,37]. In addition, some chemicals, such as toluene, can be produced from biomass components [38]. The assignments of these impurities were assigned according to the previous studies. Contamination of poplar by THF appeared at 2977 and 2875 cm^{-1} for its C-H stretching, 1063 cm^{-1} for ring deformation, and 912 cm^{-1} for CH_2 twisting [39]. The bands of dioxane were observed at 2960, 2890, 1457, 1322, 1255, 1119, 1057, 889 and 872 cm^{-1} to show its equatorial (higher frequency) C-H stretching, axial (lower frequency) C-H stretching, symmetric CH_2 deformation, CH_2 wagging, CH_2 twisting, C-O-C symmetric stretching, ring trigonal deformation, C-C stretching, and C-O-C stretching, respectively [40]. The addition of toluene on poplar caused three bands including 3069 cm^{-1} for C-H stretching, 1497 cm^{-1} for C-C stretching, and 728 cm^{-1} for C-H out of plane bending [41]. Glycerol on poplar had the bands for O-H stretching at 3341 cm^{-1}, C-H stretching at 2948 and 2897 cm^{-1}, C-H deformation of secondary alcohol at 1333 and 1239 cm^{-1}, C-O stretching of primary alcohol at 1034 cm^{-1}, and O-H bending at 923 cm^{-1} [42,43]. Chloroform contaminants also showed at 1220 and 755 cm^{-1} for C-H bending and CCl_3 stretching [34]. Pyridine contamination resulted in additional bands for C-H stretching at 3036 cm^{-1}, C-C bonding at 1583 cm^{-1}, C-N stretching at1485 cm^{-1}, C-H in plane wagging at 1438 cm^{-1}, symmetric C-H wagging at 1203 cm^{-1}, C-H wagging at 1069 cm^{-1}, C-C in plane wagging at 1032 cm^{-1}, C-H out of plane bending at 750 and 693 cm^{-1} [44,45]. The intensities of the bands at 3353 and 1653 cm^{-1} decreased with the contaminants that do not contain OH groups such as acetone, THF, dioxane, toluene, chloroform, and pyridine due to the displacement of moisture in biomass by these solvents. On the other hand, the intensity increased with the contaminants having OH groups such as water, ethanol, methanol, and glycerol.

Table 2. FTIR band assignments of common biomass processing solvents on poplar [32–45].

Contaminants	Wavenumber [cm^{-1}]	Assignments
Water	3354	O-H stretching
	1653	O-H-O scissors bending
Acetone	3005	CH_3 stretching
	2908	CH_3 stretching
	1713	C=O stretching
	1431	CH_3 deforming
	1364	CH_3 deforming
	1222	C-C stretching
Ethanol	3350	O-H stretching
	2980	C-H stretching
	1056	C-O stretching

Table 2. *Cont.*

Contaminants	Wavenumber [cm^{-1}]	Assignments
Methanol	3352	O-H stretching
	2952	C-H asymmetric stretching
	2879	C-H symmetric stretching
	1465	C-H asymmetric bending
	1450	C-H symmetric bending
	1336	O-H bending
	1068	CH$_3$ rocking
	1026	C-O stretching
Tetrahydrofuran	2977	C-H stretching
	2875	C-H stretching
	1063	Ring deformation
	912	CH$_2$ twisting
Toluene	3069	C-H stretching
	1497	C-C stretching
	728	C-H out of plane bending
Dioxane	2960	C-H stretching
	2890	C-H stretching
	1457	Symmetric CH$_2$ deformation
	1322	CH$_2$ wagging
	1255	CH$_2$ twisting
	1119	C-O-C symmetric stretching
	1057	Ring trigonal deformation
	889	C-C stretching
	872	C-O-C stretching
Glycerol	3341	O-H bending
	2948	C-H stretching
	2897	C-H stretching
	1333	C-H deformation
	1239	C-H deformation
	1034	C-O stretching
	923	O-H bending
Chloroform	1220	C-H bending
	755	CCl$_3$ stretching
Pyridine	3036	C-H stretching
	1583	C-C bonding
	1485	C-N stretching
	1438	C-H in plane wagging
	1203	Symmetric C-H wagging
	1069	C-H wagging
	1032	C-C in plane wagging
	750/693	C-H out of plane bending

3.3. Acids and Alkalis

Sulfuric acid, hydrochloric acid, acetic acid, phosphoric acid, ammonium hydroxide, and sodium hydroxide on the poplar sample were observed. As Table 3 and Figure S2 present, most acids on poplar, including sulfuric acid, hydrochloric acid, and phosphoric acid, commonly had the bands at 3370 and 1660 cm^{-1} to represent O-H bonding and O-H-O scissors bending, respectively, because of water content. The contamination bands from sulfuric acid in the literature at 1362 and 750 cm^{-1} for S=O (1362 cm^{-1}) and S-O stretching (750 cm^{-1}) were not clearly appeared in this study [46]. A relatively low concentration of sulfuric acid (4%) could be the reason for weak intensities of the contaminant. Hydrochloric acid showed H-Cl stretching at 2942 cm^{-1}, while phosphoric acid had a P-OH bond and P=O stretching at 2904 and 1161 cm^{-1}, respectively [34,47,48]. Acetic acid bands appeared at 3351, 2916, 1706, 1427, 1234, and 1031 cm^{-1} to indicate its O-H stretching, symmetric CH$_3$ stretching, C=O stretching, CH$_3$ deformation, O-H bending and CH$_3$ rocking, respectively [34]. Sodium hydroxide had the bands caused by water at 3360 and 1660 cm^{-1}, but there were no other clear contamination bands observed. Similarly, ammonium hydroxide had the IR bands at 3350 and 1660 cm^{-1} from both water and NH_3 content but N-H stretching of NH_4^+ also appeared at 2914 cm^{-1}. Previous study also said that adsorption of ammonia increased the overall polarity and resulted in the absorbance of several bands (e.g., 1115 and 1036 cm^{-1} in this study) not from the N-H vibrations [49].

Table 3. FTIR band assignments of acids and alkalis contaminants on poplar [34,46–49].

Contaminants	Wavenumber [cm^{-1}]	Assignments
	3370	O-H stretching
Sulfuric acid	1660	O-H-O scissors bending
	1362	S=O stretching
	750	S-O stretching
	3370	O-H stretching
Hydrochloric acid	2905	H-Cl stretching
	1660	O-H-O scissors bending
	3370	O-H stretching
	2905	P-OH bond
Phosphoric acid	1660	O-H-O scissors bending
	1161	P=O stretching
	1031	P=O stretching
	3351	O-H stretching
	2916	Symmetric CH_3 stretching
	1706	C=O stretching
Acetic acid	1427	CH_3 deformation
	1234	O-H bending
	1031	CH_3 rocking
	3350	N-H stretching & O-H stretching
Ammonium hydroxide	2914	N-H stretching
	1660	O-H-O scissors bending

3.4. Ionic Liquids

Besides the aforementioned chemicals, FTIR spectra and the band assignments of ionic liquids, enzymes, and biomass-derived chemicals on poplar are presented in Table 4 and Figures S3–S5. The bands from 1-butyl-3-methylimidazolium chloride contaminant were observed at 3341, 1658, and 1604 cm^{-1} representing the formation of quaternary amine salt formation with chlorine, C=C stretching, and C=N stretching, respectively. However, the band at 835 cm^{-1} representing C-N stretching vibration was not clearly observed [50]. The bands from 1-benzyl-3-methylimidazolium chloride were 2961, 1574, 765, and 633 cm^{-1} from C-H stretching, C-C stretching of ring vibration, and C-N/C-Cl in-plane bending, respectively [51]. Moreover, two bands at 1383 and 1176 cm^{-1} were observed; however, further study is needed to identify them.

Table 4. FTIR band assignments of ionic liquids, enzymes, and biomass-derived chemicals on poplar [49–65].

Contaminants	Wavenumber [cm^{-1}]	Assignments
	3341	Quaternary amine salt formation
1-Butyl-3-methylimidazolium chloride	1658	C=C stretching
	1604	C=N stretching
	2961	C-H stretching
1-Benzyl-3-methylimidazolium chloride	1574	C-C stretching ring vibration
	633	C-N/C-Cl in-plane bending

Table 4. *Cont.*

Contaminants	Wavenumber [cm^{-1}]	Assignments
ChCl-Urea	3435	NH$_2$ asymmetric stretching
	3340	NH$_2$ symmetric stretching
	1669	NH$_2$ bending vibration
	1597	OH bending vibration
	1474	CH$_3$ rocking
	1152	C-N stretching
	1062	CH$_2$ rocking
	961	Asymmetric stretching of COO
	790	C=O bonding
ChCl–PHA	3180	O-H stretching
	1681	C=O stretching
	1581	Asymmetric stretching of COO
	1282	C-O stretching vibration
	1082	C-O stretching
	953	C-N stretching
	861	CH$_2$ rocking vibrations
	838	Aromatic C-H out-of-plane bending
	786	C–C stretching
ChCl–PB	3122	The stretching vibration of the phenolic O-H
	1667	The stretching vibration of carbonyl group
	1272	Methylene
	1030	C-H binding
ChCl-PCA	3126	Bending vibration of –NH$_2$
	1675	C=O stretch of carboxylic acid
	1606	C=C stretching
	1160	C-OH stretching
	848	C-H stretching
	771	Stretching of the -OH group
Cellulase	3353	N-H stretching & O-H stretching
	2942	C-H stretching (asymmetric)
	2900	C-H stretching (symmetric)
	1642	NH$_2$ scissoring & C=N vibration
	1334	C-N stretching
	1036	C-N stretching
β-glucosidase	3351	N-H stretching
	1646	N-H bonding & C=O stretching
	1432	N-H stretching
	620	N-H out of plane bending
HMF	3364	O-H stretching
	1661	Carbonyl stretching
	1561	C=C stretching (furan ring)
Furfural	3134	C-H stretching
	2859	C-H vibration of aldehyde group
	1671	C=O in conjugated carbonyl group
	1465	C=C stretching of furan ring
	1276/1021	C-O stretching vibration
	928/884/755	C-H bending out of plane peaks

The bands of choline chloride-urea, which is a well-known DES, were at 3435 and 3340 cm^{-1}, which ascribed to the stretching of –NH$_2$ (asymmetric and symmetric), 1669 cm^{-1} for the bending vibration of –NH$_2$, 1597 cm^{-1} for bending vibration of -OH possibly due to the existence of water, 1474 cm^{-1} for CH$_3$ rocking, 1152 cm^{-1} for asymmetric C-N stretching, 1062 cm^{-1} for CH$_2$ rocking, 961 cm^{-1} for asymmetric stretching of CCO from choline structure and 790 cm^{-1} from C=O bonding [52,53]. Three lignin-based DESs, choline chloride–*p*-hydroxybenzoic acid (PHA), choline chloride–4-hydroxybenzaldehyde (PB), and choline chloride–*p*-coumaric acid (PCA), were mixed with poplar sample to observe the possible contamination bands. The bands of choline chloride–PHA were observed at 3180 cm^{-1} for O-H stretching, 1681 cm^{-1} for C=O stretching, 1581 cm^{-1} for the asymmetric stretch of COO vibrations, 1282 cm^{-1} for C-O stretching vibration, 1082 cm^{-1} for C-O stretching, 953 cm^{-1} for C-N stretching, 861 cm^{-1} for CH$_2$ rocking vibrations, 838 cm^{-1} for aromatic C-H out-of-plane bending, 786 cm^{-1} for C–C stretching [54,55]. The bands from choline chloride–PB were observed at 3122 cm^{-1} for the stretching vibration of the phenolic O-H group exhibiting intermolecular hydrogen bonding, 1667 cm^{-1} for the stretching vibration of carbonyl group, 1272 cm^{-1} for the methylene, 1030 cm^{-1} for C-H binding vibration [56]. The bands from choline chloride—PCA DES were observed at 3126 cm^{-1}, 1675 cm^{-1}, 1606 cm^{-1}, 1160 cm^{-1}, 848 cm^{-1} from bending vibration of –NH$_2$, C=O stretch of carboxylic acid, C=C stretching, C-OH stretching, C-H stretching and 771cm^{-1} from stretching of the -OH group on the second carbon of the choline chloride [57–59].

3.5. Enzymes

Enzymes such as cellulase and β-glucosidase break polysaccharides in biomass to fermentable sugars. The bands at 3353, 2942, 2900, 1642, 1334, and 1036 cm^{-1} were observed from cellulase (Table 4 and Figure S4). The bands at 3353, 2942, and 2900 cm^{-1} were from N-H/O-H stretching and the C-H stretching (asymmetric and symmetric) of cellulase. The bands at 1642, 1432, 1334, and 1036 cm^{-1} were possibly from NH$_2$ scissoring, C-C stretching, C-N stretching, and C-O stretching, respectively [60–62]. β-glucosidase also showed similar bands at 3351, 1646, 1432, and 620 cm^{-1}, which represented N-H stretching, N-H bonding and C=O stretching, N-H bending, and N-H out of plane bending, respectively [63].

3.6. Biomass-Derived Chemicals

Biomass can be contaminated by its decomposed fractions. For instance, furan-based chemicals such as furfural and hydroxymethylfurfural can be produced through the dehydration of hexoses and pentoses in biomass. As Figure S5 presents, HMF contamination showed at 3364, 1661, and 1561 cm^{-1} from O-H stretching, C=O stretching (carbonyl), and C=C stretching of furan ring, respectively [64]. Furfural also showed bands at 3134 cm^{-1} from C-H stretching of furan ring, at 2859 cm^{-1} from the C-H vibration of aldehyde group, 1671 cm^{-1} from C=O in the conjugated carbonyl group, 1465 cm^{-1} from C=C stretching of furan ring, 1276 and 1021 cm^{-1} from C-O stretching vibration, 928, 884, and 755 cm^{-1} from C-H bending out of plane peaks [65,66].

4. Conclusions

The identification of contaminants on the biomass surface after preprocessing is important to avoid the unwanted misleading of analysis data. This study investigated and discussed diagnostic FTIR bands from 26 potential chemicals, including organic solvents, acids and alkalis, ionic liquids, enzymes, and biomass-derived components through diverse biomass preprocessing. The observation of these contaminants will improve the FTIR analysis with diverse biomass and bioproducts in the biorefinery.

Supplementary Materials: The following are available online at http://www.mdpi.com/2076-3417/10/12/4345/s1, Figure S1: FTIR spectra of preprocessing solvent contaminants on poplar, Figure S2: FTIR spectra of preprocessing acid and alkaline contaminants on poplar, Figure S3: FTIR spectra of ionic liquid contaminants on poplar, Figure S4: FTIR spectra of enzyme contaminants on poplar, Figure S5: FTIR spectra of biomass-derived chemical contaminants on poplar.

Author Contributions: C.G.Y. and A.J.R. conceived and designed the research. J.Z., C.G.Y., M.L., and Y.P. carried out the experiment. J.Z., M.L., and C.G.Y. wrote the manuscript. All the authors discussed data and revised the paper. All authors have given approval to the final version of the manuscript.

Funding: This research received no external funding.

Acknowledgments: This manuscript has been authored by UT-Battelle, LLC under Contract No. DE-AC05-00OR22725 with the U.S. Department of Energy. This study was supported and performed as part of the BioEnergy Science Center (BESC) and Center for Bioenergy Innovation (CBI). The BESC and CBI are U.S. Department of Energy Bioenergy Research Centers supported by the Office of Biological and Environmental Research in the DOE Office of Science. The United States Government retains and the publisher, by accepting the article for publication, acknowledges that the United States Government retains a non-exclusive, paid-up, irrevocable, world-wide license to publish or reproduce the published form of this manuscript, or allow others to do so, for the United States Government purposes. The Department of Energy will provide public access to these results of federally sponsored research in accordance with the DOE Public Access Plan (http://energy.gov/downloads/doe-public-access-plan). The views and opinions of the authors expressed herein do not necessarily state or reflect those of the United States Government or any agency thereof. Neither the United States Government nor any agency thereof, nor any of their employees, makes any warranty, expressed or implied, or assumes any legal liability or responsibility for the accuracy, completeness, or usefulness of any information, apparatus, product, or process disclosed, or represents that its use would not infringe privately owned rights. The views and opinions of the authors expressed herein do not necessarily state or reflect those of the United States Government or any agency thereof. Neither the United States Government nor any agency thereof, nor any of their employees, makes any warranty, expressed or implied, or assumes any legal liability or responsibility for the accuracy, completeness, or usefulness of any information, apparatus, product, or process disclosed, or represents that its use would not infringe privately owned rights.

Conflicts of Interest: There are no conflicts to declare.

References

1. Pu, Y.; Cao, S.; Ragauskas, A.J. Application of quantitative 31P NMR in biomass lignin and biofuel precursors characterization. *Energy Environ. Sci.* **2011**, *4*, 3154–3166. [CrossRef]
2. Yoo, C.G.; Pu, Y.; Li, M.; Ragauskas, A.J. Elucidating Structural Characteristics of Biomass using Solution-State 2D NMR with a Mixture of Deuterated Dimethylsulfoxide and Hexamethylphosphoramide. *ChemSusChem* **2016**, *9*, 1090–1095. [CrossRef] [PubMed]
3. Yoo, C.G.; Yang, Y.; Pu, Y.; Meng, X.; Muchero, W.; Yee, K.L.; Thompson, O.A.; Rodriguez, M.; Bali, G.; Engle, N.L. Insights of biomass recalcitrance in natural Populus trichocarpa variants for biomass conversion. *Green Chem.* **2017**, *19*, 5467–5478. [CrossRef]
4. Sluiter, A.; Hames, B.; Ruiz, R.; Scarlata, C.; Sluiter, J.; Templeton, D.; Crocker, D. Determination of structural carbohydrates and lignin in biomass. *Lab. Anal. Proced.* **2010**, *1617*, 1–16.
5. Jung, S.; Foston, M.; Kalluri, U.C.; Tuskan, G.A.; Ragauskas, A.J. 3D chemical image using TOF-SIMS revealing the biopolymer component spatial and lateral distributions in biomass. *Angew. Chem.* **2012**, *124*, 12171–12174. [CrossRef]
6. Tolbert, A.K.; Yoo, C.G.; Ragauskas, A.J. Understanding the Changes to Biomass Surface Characteristics after Ammonia and Organosolv Pretreatments by Using Time-of-Flight Secondary-Ion Mass Spectrometry (TOF-SIMS). *ChemPlusChem* **2017**, *82*, 686–690. [CrossRef]
7. Sannigrahi, P.; Kim, D.H.; Jung, S.; Ragauskas, A. Pseudo-lignin and pretreatment chemistry. *Energy Environ. Sci.* **2011**, *4*, 1306–1310. [CrossRef]
8. Figueira, M.; Volesky, B.; Mathieu, H. Instrumental analysis study of iron species biosorption by Sargassum biomass. *Environ. Sci. Technol.* **1999**, *33*, 1840–1846. [CrossRef]
9. Kok, M.V.; Özgür, E. Thermal analysis and kinetics of biomass samples. *Fuel Process. Technol.* **2013**, *106*, 739–743. [CrossRef]
10. Pu, Y.; Meng, X.; Yoo, C.G.; Li, M.; Ragauskas, A.J. Analytical methods for biomass characterization during pretreatment and bioconversion. In *Valorization of Lignocellulosic Biomass in a Biorefinery: From Logistics to Environmental and Performance Impact*; Kumar, R., Ed.; Nova Science Publishers: New York, NY, USA, 2016; pp. 37–78.
11. Acquah, G.E.; Via, B.K.; Fasina, O.O.; Eckhardt, L.G. Rapid quantitative analysis of forest biomass using fourier transform infrared spectroscopy and partial least squares regression. *J. Anal. Methods Chem.* **2016**, *2016*, 1839598. [CrossRef]
12. Perkins, W. Fourier transform infrared spectroscopy. Part II. Advantages of FT-IR. *J. Chem. Educ.* **1987**, *64*, A269. [CrossRef]
13. Di Fidio, N.; Raspolli Galletti, A.M.; Fulignati, S.; Licursi, D.; Liuzzi, F.; De Bari, I.; Antonetti, C. Multi-Step Exploitation of Raw *Arundo donax* L. for the Selective Synthesis of Second-Generation Sugars by Chemical and Biological Route. *Catalysts* **2020**, *10*, 79. [CrossRef]
14. Licursi, D.; Antonetti, C.; Fulignati, S.; Corsini, A.; Boschi, N.; Galletti, A.M.R. Smart valorization of waste biomass: Exhausted lemon peels, coffee silverskins and paper wastes for the production of levulinic acid. *Chem. Eng. Trans.* **2018**, *65*, 637.
15. Lara-Serrano, M.; Morales-delaRosa, S.; Campos-Martín, J.M.; Fierro, J.L.G. Fractionation of Lignocellulosic Biomass by Selective Precipitation from Ionic Liquid Dissolution. *Appl. Sci.* **2019**, *9*, 1862. [CrossRef]
16. Shen, X.J.; Wen, J.L.; Mei, Q.Q.; Chen, X.; Sun, D.; Yuan, T.Q.; Sun, R.C. Facile fractionation of lignocelluloses by biomass-derived deep eutectic solvent (DES) pretreatment for cellulose enzymatic hydrolysis and lignin valorization. *Green Chem.* **2019**, *21*, 275–283. [CrossRef]

17. Agbor, V.B.; Cicek, N.; Sparling, R.; Berlin, A.; Levin, D.B. Biomass pretreatment: Fundamentals toward application. *Biotechnol. Adv.* **2011**, *29*, 675–685. [CrossRef] [PubMed]

18. Yoo, C.G.; Pu, Y.; Ragauskas, A.J. Ionic liquids: Promising green solvents for lignocellulosic biomass utilization. *Curr. Opin. Green Sustain. Chem.* **2017**, *5*, 5–11. [CrossRef]

19. Nguyen, T.Y.; Cai, C.M.; Osman, O.; Kumar, R.; Wyman, C.E. CELF pretreatment of corn stover boosts ethanol titers and yields from high solids SSF with low enzyme loadings. *Green Chem.* **2016**, *18*, 1581–1589. [CrossRef]

20. Yang, B.; Dai, Z.; Ding, S.-Y.; Wyman, C.E. Enzymatic hydrolysis of cellulosic biomass. *Biofuels* **2011**, *2*, 421–449. [CrossRef]

21. Rasmussen, H.; Sørensen, H.R.; Meyer, A.S. Formation of degradation compounds from lignocellulosic biomass in the biorefinery: Sugar reaction mechanisms. *Carbohydr. Res.* **2014**, *385*, 45–57. [CrossRef]

22. Li, M.; Pu, Y.; Yoo, C.G.; Gjersing, E.; Decker, S.R.; Doeppke, C.; Shollenberger, T.; Tschaplinski, T.J.; Engle, N.L.; Sykes, R.W. Study of traits and recalcitrance reduction of field-grown COMT down-regulated switchgrass. *Biotechnol. Biofuels* **2017**, *10*, 12. [CrossRef] [PubMed]

23. Faix, O. Fourier transform infrared spectroscopy. In *Methods in Lignin Chemistry*; Lin, S.Y., Dence, C.W., Eds.; Springer: Berlin/Heidelberg, Germany, 1992; pp. 83–109.

24. Sim, S.F.; Mohamed, M.; Lu, N.A.L.M.I.; Sarman, N.S.P.; Samsudin, S.N.S. Computer-assisted analysis of fourier transform infrared (FTIR) spectra for characterization of various treated and untreated agriculture biomass. *BioResources* **2012**, *7*, 5367–5380. [CrossRef]

25. Pandey, K. A study of chemical structure of soft and hardwood and wood polymers by FTIR spectroscopy. *J. Appl. Polym. Sci.* **1999**, *71*, 1969–1975. [CrossRef]

26. Ciolacu, D.; Ciolacu, F.; Popa, V.I. Amorphous cellulose—Structure and characterization. *Cell. Chem. Technol.* **2011**, *45*, 13.

27. Yang, H.; Yan, R.; Chen, H.; Lee, D.H.; Zheng, C. Characteristics of hemicellulose, cellulose and lignin pyrolysis. *Fuel* **2007**, *86*, 1781–1788. [CrossRef]

28. Le, D.M.; Nielsen, A.D.; Sørensen, H.R.; Meyer, A.S. Characterisation of Authentic Lignin Biorefinery Samples by Fourier Transform Infrared Spectroscopy and Determination of the Chemical Formula for Lignin. *Bioenergy Res.* **2017**, *10*, 1025–1035. [CrossRef]

29. Liu, C.F.; Xu, F.; Sun, J.X.; Ren, J.L.; Curling, S.; Sun, R.C.; Fowler, P.; Baird, M.S. Physicochemical characterization of cellulose from perennial ryegrass leaves (*Lolium perenne*). *Carbohydr. Res.* **2006**, *341*, 2677–2687. [CrossRef]

30. Xu, F.; Sun, J.-X.; Sun, R.; Fowler, P.; Baird, M.S. Comparative study of organosolv lignins from wheat straw. *Ind. Crop. Prod.* **2006**, *23*, 180–193. [CrossRef]

31. Sills, D.L.; Gossett, J.M. Using FTIR to predict saccharification from enzymatic hydrolysis of alkali-pretreated biomasses. *Biotechnol. Bioeng.* **2012**, *109*, 353–362. [CrossRef]

32. Mojet, B.L.; Ebbesen, S.D.; Lefferts, L. Light at the interface: The potential of attenuated total reflection infrared spectroscopy for understanding heterogeneous catalysis in water. *Chem. Soc. Rev.* **2010**, *39*, 4643–4655. [CrossRef]

33. Zhang, K.; Pei, Z.; Wang, D. Organic solvent pretreatment of lignocellulosic biomass for biofuels and biochemicals: A review. *Bioresour. Technol.* **2016**, *199*, 21–33. [CrossRef] [PubMed]

34. The Virtual Planetary Laboratory Molecular Database. Available online: http://vpl.astro.washington.edu/spectra/allmoleculeslist.htm (accessed on 8 May 2020).

35. Plyler, E.K. Infrared Spectra of Methanol, Ethanol, and *n*-Propanol. *J. Res. Natl. Bur. Stand.* **1952**, *48*, 281–286. [CrossRef]

36. Conklin, A., Jr.; Goldcamp, M.J.; Barrett, J. Determination of ethanol in gasoline by FT-IR spectroscopy. *J. Chem. Educ.* **2014**, *91*, 889–891. [CrossRef]

37. Hu, S.; Li, Y. Two-step sequential liquefaction of lignocellulosic biomass by crude glycerol for the production of polyols and polyurethane foams. *Bioresour. Technol.* **2014**, *161*, 410–415. [CrossRef] [PubMed]

38. Elfadly, A.; Zeid, I.; Yehia, F.; Abouelela, M.; Rabie, A. Production of aromatic hydrocarbons from catalytic pyrolysis of lignin over acid-activated bentonite clay. *Fuel Process. Technol.* **2017**, *163*, 1–7. [CrossRef]

39. Dwivedi, A.; Baboo, V.; Bajpai, A. Fukui Function Analysis and Optical, Electronic, and Vibrational Properties of Tetrahydrofuran and Its Derivatives: A Complete Quantum Chemical Study. *J. Theor. Chem.* **2015**, *2015*, 345234. [CrossRef]

40. Borowski, P.; Gac, W.; Pulay, P.; Woliński, K. The vibrational spectrum of 1, 4-dioxane in aqueous solution–theory and experiment. *New J. Chem.* **2016**, *40*, 7663–7670. [CrossRef]

41. IR Spectroscopy Tutorial. Available online: https://orgchemboulder.com/Spectroscopy/irtutor/tutorial.shtml (accessed on 8 May 2020).

42. Wen Yee, T.; Tin Sin, L.; Rahman, W.; Samad, A. Properties and interactions of poly (vinyl alcohol)-sago pith waste biocomposites. *J. Compos. Mater.* **2011**, *45*, 2199–2209. [CrossRef]

43. Danish, M.; Mumtaz, M.W.; Fakhar, M.; Rashid, U. Response surface methodology based optimized purification of the residual glycerol from biodiesel production process. *Chiang Mai J. Sci.* **2015**, *43*, 1570–1582.

44. Swoboda, A.; Kunze, G. Infrared study of pyridine adsorbed on montmorillonite surfaces. *Clay Clay Miner.* **1964**, *13*, 277. [CrossRef]

45. Testa, A.C. Molecular Vibrations of Pyridine. Available online: http://facpub.stjohns.edu/~{}testaa/anim27vib.html (accessed on 6 May 2020).

46. Segneanu, A.E.; Gozescu, I.; Dabici, A.; Sfirloaga, P.; Szabadai, Z. Organic compounds FT-IR spectroscopy. In *Macro To Nano Spectroscopy*; Uddin, J., Ed.; InTech: Rijeka, Croatia, 2012; pp. 145–164.

47. Eisazadeh, A.; Kassim, K.A.; Nur, H. Physicochemical characteristics of phosphoric acid stabilized bentonite. *Electron. J. Geotech. Eng.* **2010**, *15*, 327–335.

48. Arai, Y.; Sparks, D.L. ATR–FTIR spectroscopic investigation on phosphate adsorption mechanisms at the ferrihydrite–water interface. *J. Colloid Interface Sci.* **2001**, *241*, 317–326. [CrossRef]

49. Valentin, R.; Horga, R.; Bonelli, B.; Garrone, E.; Renzo, F.D.; Quignard, F. FTIR spectroscopy of NH_3 on acidic and ionotropic alginate aerogels. *Biomacromolecules* **2006**, *7*, 877–882. [CrossRef]

50. Dharaskar, S.A.; Varma, M.N.; Shende, D.Z.; Yoo, C.K.; Wasewar, K.L. Synthesis, characterization and application of 1-butyl-3 methylimidazolium chloride as green material for extractive desulfurization of liquid fuel. *Sci. World J.* **2013**, *2013*, 395274. [CrossRef] [PubMed]

51. Seethalakshmi, K.; Jasmine Vasantha Rani, E.; Padmavathy, R. Study of vibrational spectra and solvation number of non-aqueous solutions of 1-benzyl-3-dimethylimidazolium chloride through ultrasonic technique. *Int. J. Recent Sci. Res.* **2015**, *6*, 2347–2349.

52. Yue, D.; Jia, Y.; Yao, Y.; Sun, J.; Jing, Y. Structure and electrochemical behavior of ionic liquid analogue based on choline chloride and urea. *Electrochim. Acta* **2012**, *65*, 30–36. [CrossRef]

53. Du, C.; Zhao, B.; Chen, X.-B.; Birbilis, N.; Yang, H. Effect of water presence on choline chloride-2urea ionic liquid and coating platings from the hydrated ionic liquid. *Sci. Rep.* **2016**, *6*, 29225. [CrossRef]

54. Dega-Szafran, Z.; Dutkiewicz, G.; Kosturkiewicz, Z.; Szafran, M. Crystal structure and spectroscopic properties of the complex of trigonelline hydrate with *p*-hydroxybenzoic acid. *J. Mol. Struct.* **2011**, *985*, 219–226. [CrossRef]

55. Sun, R.; Tomkinson, J.; Bolton, J. Separation and characterization of lignins from the black liquor of oil palm trunk fiber pulping. *Sep. Sci. Technol.* **1999**, *34*, 3045–3058. [CrossRef]

56. Shareef, B.A.; Waheed, I.F.; Jalaot, K.K. Preparation and Analytical Properties of 4-Hydroxybenzaldehyde, Biuret and Formaldehyde Terpolymer Resin. *Orient. J. Chem.* **2014**, *29*, 1391–1397. [CrossRef]

57. Kaur, J.; Katopo, L.; Hung, A.; Ashton, J.; Kasapis, S. Combined spectroscopic, molecular docking and quantum mechanics study of β-casein and p-coumaric acid interactions following thermal treatment. *Food Chem.* **2018**, *252*, 163–170. [CrossRef] [PubMed]

58. Moosavinejad, S.M.; Madhoushi, M.; Vakili, M.; Rasouli, D. Evaluation of degradation in chemical compounds of wood in historical buildings using FT-IR and FT-Raman vibrational spectroscopy. *Maderas-Cienc. Tecnol.* **2019**, *21*. [CrossRef]

59. Asare, S. Synthesis, Characterization and Molecular Dynamic Simulations of Aqueous Choline Chloride Deep Eutectic Solvents. Ph.D. Thesis, South Dakota State University, Brookings, SD, USA, 2018.

60. Bohara, R.A.; Thorat, N.D.; Pawar, S.H. Immobilization of cellulase on functionalized cobalt ferrite nanoparticles. *Korean J. Chem. Eng.* **2016**, *33*, 216–222. [CrossRef]

61. Zhang, D.; Hegab, H.E.; Lvov, Y.; Snow, L.D.; Palmer, J. Immobilization of cellulase on a silica gel substrate modified using a 3-APTES self-assembled monolayer. *SpringerPlus* **2016**, *5*, 48. [CrossRef] [PubMed]

62. Swarnalatha, V.; Ester, R.A.; Dhamodharan, R. Immobilization of α-amylase on gum acacia stabilized magnetite nanoparticles, an easily recoverable and reusable support. *J. Mol. Catal. B-Enzym.* **2013**, *96*, 6–13. [CrossRef]

63. Bai, H.; Wang, H.; Sun, J.; Irfan, M.; Han, M.; Huang, Y.; Han, X.; Yang, Q. Production, purification and characterization of novel beta glucosidase from newly isolated Penicillium simplicissimum H-11 in submerged fermentation. *EXCLI J.* **2013**, *12*, 528.
64. Tsilomelekis, G.; Josephson, T.R.; Nikolakis, V.; Caratzoulas, S. Origin of 5-hydroxymethylfurfural stability in water/dimethyl sulfoxide mixtures. *ChemSusChem* **2014**, *7*, 117–126. [CrossRef]
65. Garba, N.A.; Muduru, I.; Sokoto, M.A.; Dangoggo, S.M. Production of liquid hydrocarbons from millet husk via catalytic hydrodeoxygenation in NIO/AL$_2$O$_3$ catalysts. In *WIT Transactions on Ecology and the Environment*; Syngellakis, S., Magaril, E., Eds.; WIT PRESS: Billerica, MA, USA, 2018; pp. 125–130.
66. Allen, G.; Bernstein, H.J. Internal rotation: VIII. The infrared and raman spectra of furfural. *Can. J. Chem.* **1955**, *33*, 1055–1061. [CrossRef]

 *applied sciences*

Article

Drying Effect on Enzymatic Hydrolysis of Cellulose Associated with Porosity and Crystallinity

Bonwook Koo [1],*, Jaemin Jo [1] and Seong-Min Cho [2]

[1] Green and Sustainable Materials R & D Department, Korea Institute of Industrial Technology, Cheonan-si 31056, Korea; jjm1234@kitech.re.kr

[2] Department of Forest Sciences, Seoul National University, Seoul 08826, Korea; csmin93@snu.ac.kr

* Correspondence: bkoo@kitech.re.kr; Tel.: +82-41-589-8409

Received: 27 July 2020; Accepted: 7 August 2020; Published: 11 August 2020

Featured Application: Drying causes irreversible structural changes in cellulose and the changes are intimately associated with porosity, including pore volume and surface area. These changes must be considered for the application of cellulose in high value industry such as sustainable polymers that use cellulose nanofiber and sustainable sugar production.

Abstract: The effect of drying on the enzymatic hydrolysis of cellulose was determined by analysis of porosity and crystallinity. Fiber hornification induced by drying produced an irreversible reduction in pore volume due to shrinkage and pore collapse, and the decrease in porosity inhibited enzymatic hydrolysis. The drying effect index (DEI) was defined as the difference in enzymatic digestibility between oven- and never-dried pulp, and it was determined that more enzymes caused a higher DEI at the initial stage of enzymatic hydrolysis and the highest DEI was also observed at the earlier stages with higher enzyme dosage. However, there was no significant difference in the DEI with less enzymes because cellulose conversion to sugars during hydrolysis did not enhance enzymatic hydrolysis due to the decrease in enzyme activity. The water retention value (WRV) and Simons' staining were used to measure pore volume and to investigate the cause of the decrease in enzymatic hydrolysis. A decrease in enzyme accessibility induced by the collapse of enzymes' accessible larger pores was determined and this decreased the enzymatic hydrolysis. However, drying once did not cause any irreversible change in the crystalline structure, thus it seems there is no correlation between enzymatic digestibility and crystalline structure.

Keywords: drying effect; cellulose; enzymatic hydrolysis; hornification; porosity

---

1. Introduction

Biorefinery platforms using biomass have been studied widely in recent years because of their low carbon footprint [1]. The U.S. Department of Energy (DOE) announced the top 12 platform chemicals that can be produced from biomass and that it would consider the market and potential application of these in industry [2,3]. Most of the platform chemicals can be produced from sugars, which indicates that low cost sugar production from biomass is important to achieve a low carbon footprint [4]. Lignocellulosic biomass has been suggested as the feedstock for sustainable sugar production and various conversion processes including pretreatment and saccharification have been studied to facilitate effective sugar production from lignocellulosic biomass [5–7].

Sustainable sugar is mainly produced by enzymatic saccharification and pretreatment must be performed to improve enzyme accessibility to cellulose [8,9]. The pretreatment exposes cellulose by partial removal of hemicellulose and lignin, and it facilitates enzyme access to cellulose. After pretreatment, the exposed cellulose is likely to be partially dried and the cellulose drying affects its enzymatic digestibility [10]. Once cellulose has been dried, the dimensions of the cellulose are changed due to the collapse of pores and shrinkage of the internal volume [11,12]. The structural change stiffens the cellulose via a process known as "hornification" [13].

The process of hornification is frequently explained as irreversible, or partially irreversible hydrogen bonding upon drying or water removal [14] due to the aggregation of the cellulose chains [15]. The cellulose chains are aggregated due to the removal of water by heating and then they are not able to fully open up to the next exposure to water [16]. The aggregation may collapse the pore structure, and thus a decrease in pore volume and surface area can be expected [12,17]. Crystallinity in cellulose fiber has been considered as one of the major properties altered by the aggregation [18]. It was reported that several cycles of drying and rewetting caused irreversible change in crystalline structure, which could not be recovered due to the growth of crystalline domains through co-crystallization [19]. However, no change in crystalline index by the recycling of paper has been reported [20].

It has been reported that pore volume is significantly decreased by the process of repetitive drying and rewetting. Repetitive recycling of delignified and alkali-extracted pulp was carried out by rewetting and drying, and it decreased the Brunauer–Emmett–Teller (BET) surface area and pore volume of cellulose, as measured by the water retention value (WRV). In a five-times recycling procedure, the BET surface area and the WRV decreased by 22.9% and 35.9%, respectively, compared with non-dried and rewet pulp [17].

In addition, in previous studies on the effect of drying, it was reported that hornification induced by drying affects enzymatic hydrolysis [21,22]. The enzymatic digestibility of dried substrate was significantly lower than that of never-dried substrate, and the reduction in enzymatic conversion was caused by a decrease in pore volume, which can be evaluated by WRV [22]. Hornification induced by drying caused the collapse of larger pores while increasing the number of smaller pores, which are not accessible to enzymes [21]. This suggests that the collapse or closure of larger pores could be a primary reason for a reduction in enzyme accessibility to cellulose, given the size of the cellulase core.

Although numerous studies have been performed to investigate the drying effect on enzymatic hydrolysis [23], little research on the effect with regard to enzyme dosage and the kinetics of hydrolysis has been performed. This study aims to investigate the drying effect on enzymatic hydrolysis at different enzyme dosages to determine the hydrolysis kinetics and the limitation of enzymatic digestibility. In addition, several properties of the cellulose structure including porosity, enzyme accessible surface area, and crystallinity were determined to investigate their effect on enzymatic hydrolysis.

2. Materials and Methods

2.1. Sample Preparation

A fully bleached and never-dried hardwood pulp was obtained from a mill in the southeast of the United States and used in this study. The pulp was washed thoroughly with plenty of tap water. Handsheets were made using the washed pulp to prevent the formation of fiber flocks, which might affect the enzymatic digestibility of pulp due to their effect on the enzyme-accessible surface area. It has been reported that fiber bundles in a wet state can easily form very strong fiber flocks of varying size upon drying due to inter- and intra-fiber hydrogen bonding [22]. The washed hardwood pulp, which weighed about 2.0 g was disintegrated in a standard disintegrator (Lorentzen & Wettre, Kista, Sweden) for 5 min with 5% solid consistency at room temperature and then diluted to 1% solid consistency with deionized water. The diluted pulp suspension was used to make handsheets according to the TAPPI Standard method, T205 sp-95 [24] using a standard laboratory handsheet mold. The weight was 2.0 g per sheet (based on the oven-dried weight) and the handsheets were stored in a cold room at 4 °C using a plastic bag to prevent biological decomposition and drying.

2.2. Compositional Analysis

The compositional analysis of the pulp was performed according to NREL (National Renewable Energy Laboratory) Standard Procedures [25]. Sulfuric acid hydrolysis w performed in two stages with 72% and 4% of sulfuric acid. First, the pulp was hydrolyzed with 72% of sulfuric acid at 30 °C for 1 hr and then the acid was diluted to 4% for the second hydrolysis at 121 °C for 1 hr. After two stages of sulfuric acid hydrolysis, the hydrolysate from the pulp was filtrated by a 0.2 μm Milipore filter and analyzed by high-performance liquid chromatography (Agilent 1200; Agilent Technologies, Palo Alto, CA, USA) to measure the amount of structural sugars. The sugars were separated with a Shodex SP0810 column at a temperature of 80 °C with a flow rate of 0.5 mL/min using deionized water as an eluent. A refractive index detector was used to quantify the arabinose, galactose, glucose, xylose, and mannose in the hydrolysates. The chemical composition of the pulp is shown in Table 1.

Table 1. Chemical composition of the bleached hardwood pulp.

Carbohydrates						Lignin			Ash	Total Mass Balance (%)
Glu	Xyl	Gal	Ara	Man	Total	KL	ASL	Total		
79.5	16.4	ND	ND	ND	95.9	0.4	0.5	0.9	0.3	97.1

All values were calculated based on the oven-dried weight of pulp; Glu: glucan; Xyl: xylan; Man: mannan; Gal: galactan; Ara: arabinan; KL: Klason lignin; ASL: acid soluble lignin; ND: not detected.

2.3. Drying and Rewetting of the Pulp

The handsheets were subject to drying and rewet to induce fiber hornification. Two different drying methods, oven and freeze drying, were applied to determine the drying effect on enzymatic hydrolysis. Oven drying of the handsheets was conducted in a drying oven at 105 °C for 24 h and freeze drying was conducted at −40 °C in a freeze dryer for 48 h. After each drying, the dried handsheets were stored in plastic bags for enzymatic hydrolysis and analysis of those properties. Rewetting the dried pulp was performed by disintegration for 5 min at 5% of consistency and then the water was drained out by centrifuge. Rewetted pulp fibers were also stored in a plastic bag and used for enzymatic hydrolysis and analysis of those properties.

2.4. Enzymatic Hydrolysis

Commercial cellulase including β-glucosidase (C-tec2) and a xylanase (H-tec2) provided by Novozymes (Franklinton, NC, USA), were used for enzymatic hydrolysis. Cellulase dosages of 4, 10 and 20 FPU per g of pulp (5.2, 13.3 and 26.7 enzyme protein mg per g of pulp) and a ninth of the xylanase were used to determine the drying effect depending on enzyme dosage. Enzymatic hydrolysis was performed in 20 mL of 100 mM acetate buffer (pH 5.0) at a 5% (*w/v*) solids loading. The substrates and enzymes were incubated in a shaker at 50 °C and 180 rpm for 120 h. The enzymatic digestibility was determined in duplicate by the amount of sugars released during enzymatic hydrolysis based on the amount of structural sugars in the original pulp. The amount of released sugars in the enzymatic hydrolysate was quantified by the HPLC (high pressure liquid chromatography) [26].

2.5. Analysis of Pulp Properties

2.5.1. Porosity by Water Retention Value and Simons' Staining Method

When pulp fibers are dried, internal fiber volume shrinks [11] and the shrinkage prevents the accurate measurement of porosity due to the pore collapse caused by drying. Thus, the measurement needs to be done in a wet state [27] and two independent methods, water retention value (WRV) and Simon's staining method were selected to evaluate the porosity of pulp in wet state.

The WRV has been used to estimate the fiber saturation point, which correlates to the amount of water in the cell wall pores or the volume of pores, and it has been found that the drying effect induced by fiber hornification is reflected in a reduction in the WRV [22]. The WRV was determined using the

TAPPI Useful method with a centrifugal force (Eppendorf North America, Hauppauge, NY, USA) of 900 rcf (2400 rpm) for 30 min [28]. The centrifuged wet pulp was first weighed and then oven dried at 105 °C overnight. The WRV was calculated by the percentage of water retained in the dried pulp, i.e.,

$$\text{WRV (\%)} = [(W_{wet} - W_{dried})/W_{dried}] * 100 \tag{1}$$

in which:

- W_{wet}: weight of wet pulp
- W_{dried}: weight of dried pulp

In addition to the WRV, the Simon's staining method was performed to analyze the porosity, including the enzyme accessible surface area of the pulp, depending on the drying method. For the Simons' staining method, the 1:1 staining method were used in this study [29]. Blue dye (DB) and orange dye (DO) were dissolved in nanopure water for the preparation of each dye solution and the final concentrations of DB and DO were 1% (*w/v*), respectively. For the orange dye, only higher molecular fractions were used after the fractionation. The dye solution was then prepared with 1 mL of the DB solution, 1 mL of the DO solution, 10 mL of 10% (*w/v*) NaCl and 70 mL of nanopure water. For staining, a 25 mg of sample (dry weight) in the dye solution was incubated in a 75 °C shaking incubator at 200 rpm for 48 h. The stained samples were then filtered, rinsed with a minimum amount of distilled water and stripped with 25% (*w/v*) aqueous pyridine at 45 °C for 18 h. The dye stripping solution was then analyzed using a UV-Vis spectrophotometer to determine the concentration of DO and DB (the maximum absorbance of DB and DO was at 624 and 455 nm, respectively). The concentration of DO and DB dyes in the dye stripping solution (C_O and C_B) was determined using the following two equations, Equations (2) and (3) (using the Lambert–Beer law for a binary mixture), which were solved simultaneously [30].

$$A_{455nm} = \varepsilon_{O/455} LC_O + \varepsilon_{B/455} LC_B \tag{2}$$

$$A_{624nm} = \varepsilon_{O/624} LC_O + \varepsilon_{B/624} LC_B \tag{3}$$

in which:

- A: absorbance
- ε: absorptivity
- $\varepsilon_{O/455} = 35.62$, $\varepsilon_{B/455} = 2.59$, $\varepsilon_{O/624} = 0.19$, $\varepsilon_{B/624} = 15.62$ L·g^{-1}·cm^{-1} and L = 1 cm

2.5.2. Crystallinity by ^{13}C CPMAS Solid-State NMR Analysis

High-resolution ^{13}C CPMAS NMR spectra were collected by a Bruker Avance 200 MHz spectrometer (Bruker BioSpin Corporation, Billerica, MA, USA) with a 7 mm probe, operating at 50.13 MHz for ^{13}C, at room temperature. The spinning speed was 7000 Hz, contact pulse 2 ms, acquisition time 51.3 ms and delay between pulses was 4 s, with 20,000 scans. The adamantine peak was used as an external reference (δC 38.3 ppm). The samples were hydrated with deionized water (ca. 43%) before recording the spectra. The ^{13}C chemical shifts were given in δ values (ppm) and each peak was assigned according to the values in the literature [19]: C6-amorphous (61.3 to 62.1 ppm), C6-crystalline (65.2 to 65.8 ppm), C2, C3 and C5 (71.2 to 75.6 ppm), C4-amorphous (83.5 to 84.4 ppm), C4-crystalline (88.2 to 89.6 ppm), and C1 (104.2 to 107.8 ppm).

3. Results and Discussion

3.1. Drygin Effect on Enzymatic Hydrolysis

The effect of drying on enzymatic hydrolysis was evaluated by enzymatic digestibility, which was determined by the amount of sugars released during enzymatic hydrolysis with 10 FPU/g-pulp of cellulase. It was calculated based on the amount of structural sugars in the original pulp.

Compared with oven-dried (OD) pulp, the enzymatic digestibility of never-dried (ND) pulp increased rapidly in the initial stage of the hydrolysis, but the hydrolysis rate of the ND pulp reduced gradually as the hydrolysis proceeded (Figure 1b). A decrease in hydrolysis rate has been ascribed to several factors such as the transformation of cellulose into a structurally resistant form, inhibition of enzyme action by accumulated products, and the completeness of hydrolysis [31]. Here, it seems that the decrease in hydrolysis rate of the ND pulp can be attributed to the completeness of the hydrolysis because the digestibility leveled off at 90%. Though the enzymatic digestibility of freeze-dried (FD) pulp was slightly lower than that of the ND pulp, a similar trend was observed in the hydrolysis of the FD pulp. This suggested that freeze drying caused mild hornification, unlike oven drying [21].

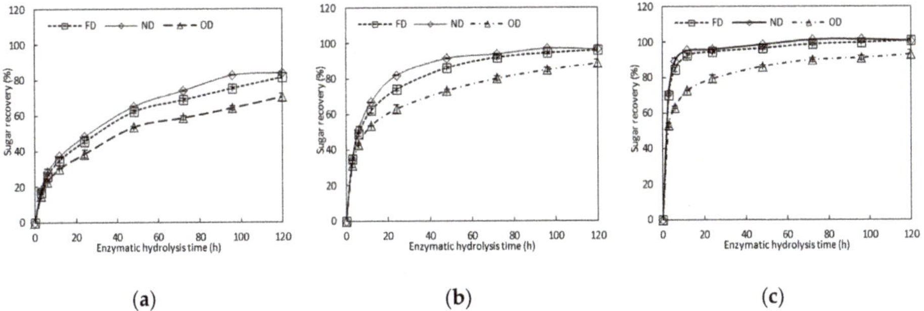

Figure 1. Enzymatic digestibility (sugar recovery) depending on the enzyme dosage of (**a**) 4 FPU/g biomass, (**b**) 10 FPU/g biomass, and (**c**) 20 FPU/g biomass.

Enzymatic hydrolysis of the ND pulp showed the highest digestibility (96.6%) when the hydrolysis was carried out with 10 FPU of cellulase for 120 h. However, the OD pulp showed significantly low digestibility though that of the FD pulp was almost similar to that of the ND pulp (96.1%) (Figure 1b). Pore volume has been considered as one of the most important properties affecting enzymatic hydrolysis [32,33] and fiber hornification induced by drying produces irreversible reduction in pore volume due to shrinkage and collapse of pores [22,32]. The reduction in pore volume decreases enzyme accessibility [22] and it reduces enzymatic digestibility due to less enzyme accessible surface area [32]. Therefore, the difference in enzymatic digestibility depending on drying, as shown in Figure 1b, should be caused by the reduction in pore volume induced by drying. However, there was a significant difference in the enzymatic digestibility of the FD and OD pulp. Again, it was considered that the better hydrolyzability of the FD pulp compared to the OD pulp, was due to the milder hornification induced by freeze drying [21]. The mild hornification can be explained by the rigidity of fibers at initial freezing [34]. The freezing gave a sort of rigidity to the fibers and this maintained the pore structure, thus minimizing pore collapse during drying. Therefore, enzymatic hydrolysis of the FD pulp showed better hydrolyzability compared to the OD pulp.

The difference in digestibility between the ND and OD pulp decreased gradually as the hydrolysis progressed because the digestibility of the OD pulp kept increasing while that of the ND pulp had already leveled off. The OD pulp had less pore volume due to fiber hornification, and thus there was a big difference in enzymatic digestibility compared to the ND pulp in the initial stages [17]. However new pores were formed by cellulose conversion to sugars during hydrolysis and this can give enzyme the space to access remaining cellulose [35]. Thus, the difference in digestibility between the ND and OD pulps was reduced gradually as the hydrolysis progressed. It is considered that the lower enzymatic digestibility of the OD pulp at the initial stage was induced by pore collapse; however, it can be recovered at a later stage because the pore volume increased due to cellulose conversion to sugars during enzymatic hydrolysis, even though hornification leads to irreversible or partial irreversible collapse of pore and it causes the difference in digestibility at the final stage of hydrolysis [32]. In order to determine the kinetics of enzymatic hydrolysis, the Langmuir equation style fitting was used, which enabled the prediction of the final digestibility after 120 h of enzymatic hydrolysis. The Langmuir equation style for the determination is as follows:

$$S_t = (S_{max} \cdot t)/(k + t) \tag{4}$$

where S_t is enzymatic digestibility at a time, S_{max} is the maximum digestibility predicted (%), k is the Langmuir style equilibrium constant and t is time (h). This equation can be re-organized by the Lineweaver–Burk method to linearize the rate of expression as follows:

$$1/S_t = k \cdot (1/S_{max}) \cdot (1/t) + 1/S_{max} \tag{5}$$

The kinetic parameters were inferred with an intercept at $1/S_{max}$ and a slope of $k \cdot (1/S_{max})$ and are summarized in Table 2. The predicted final digestibility of the OD pulp with 10 FPU was 82.6%, which was less than that of the ND pulp (102.0%) (Table 2). It was proved that there was a big difference between the ND and OD pulps with regard to enzyme accessibility. The difference was caused by the irreversible transformation of cellulose into a structurally resistant form by hornification and the inhibition of enzyme action by accumulated products [31]. All of the predicted digestibilities depending on enzyme dosage and drying, were similar to the real values.

Table 2. The prediction of final enzymatic digestibility depending on enzyme dosage and drying.

	Never Dried			Freeze Dried			Oven Dried		
FPU/g-Pulp	4	10	20	4	10	20	4	10	20
S_{max}	78.1	102.0	103.1	74.1	95.2	101.0	62.9	82.6	90.1
K	10.1	5.9	1.2	10.5	5.4	1.3	10.0	5.1	2.2

3.2. Drying Effect Depending on Enzyme Dosage

The drying effect was determined depending on enzyme dosage. Enzymatic hydrolysis was performed with 4 FPU and 20 FPU of cellulase and the enzymatic digestibility after 120 h was predicted using the same method as above with 10 FPU (Table 2).

The predicted enzymatic digestibility after 120 h increased as the enzyme dosage increased, regardless of drying. The increase in the rate of digestibility was more remarkable in the hydrolysis of the OD pulp, and the rate increased by 20.0 (ND), 24.1 (FD) and 32.3% (OD) when the enzyme dosage increased from 4 to 20 FPU. It is believed that a higher enzyme dosage can convert more cellulose to sugars in a short period of enzymatic hydrolysis, and thus the digestibility of the OD pulp increases significantly as the enzyme dosage is increased. There was a 26.0% increase in the digestibility of the OD pulp when the enzyme dosage was increased from 4 to 10 FPU, which was much higher than the 5.1% increase in digestibility when the enzymes were increased from 10 to 20 FPU. This suggested that an enzyme dosage 10 FPU was enough to improve enzymatic digestibility by pore formation during hydrolysis.

The kinetics of enzymatic hydrolysis with 4 FPU and 20 FPU are shown in Figure 1a,c. There was no leveling off on the enzymatic hydrolysis with 4 FPU. Enzymatic hydrolysis with lower enzyme dosages requires longer hydrolysis time to obtain maximum enzymatic digestibility [36], and 120 h of hydrolysis time might not be enough to achieve maximum digestibility with 4 FPU of enzyme dosage. However, the predicted digestibility with 4 FPU was 62.9%, which was slightly lower than the actual digestibility of 70.4% after 120 h (Tables 2 and 3). It was considered that hydrolysis with 4 FPU progressed steadily over 120 h and it had already reached the maximum digestibility even though it did not appear to level off. There was no further increase in digestibility with 4 FPU as a result of enzyme inhibition caused by accumulated products such as released sugars [37]. When 20 FPU was used, the enzymatic digestibility of the ND and FD pulps leveled off in 24 h and both final digestibilities were almost 100% after 120 h. Although the digestibility of the OD pulp also increased when the enzyme dosage increased to 20 FPU, it was still lower than that of the ND and FD pulp due to the drying effect induced by fiber hornification. The enzymatic digestibility after 120 h and the predicted final digestibility of the OD pulp with 20 FPU was 93.2% and 90.1%, respectively. It indicated that the drying effect on enzymatic hydrolysis can vary depending on enzyme dosage, but there is likely to be a difference in the final enzymatic digestibility of the ND and OD pulps due to irreversible structural change in the cellulose [38]. The drying effect described by the differences in the enzymatic digestibility of the ND and OD pulp with 20 FPU was clearer at the initial stage and it was different compared to the drying effect with 4 FPU, which did not show any big differences as the enzymatic hydrolysis progressed. Thus, the drying effect can vary depending on enzyme dosage and the term "Drying effect index" (DEI) was suggested to define the drying effect. The equation for calculating the DEI is as follows (Equation (6)), and the DEI was plotted, as shown in Figure 2.

$$DEI\ (\%) = 100 - (Digestibility\ of\ OD\ pulp/Digestibility\ of\ ND\ pulp) * 100 \tag{6}$$

The highest DEI, which represents the biggest drying effect, increased as enzyme dosage increased and were revealed at the earlier stages of hydrolysis (Figure 2). The highest DEIs were 22.3 (4 FPU), 24.1 (10 FPU) and 29.3 (20 FPU) and these were observed at 96, 24 and 6 h, respectively. Enzymatic hydrolysis of the OD pulp began in the non-collapsed part first, followed by hydrolysis of the collapsed part because enzyme access to the non-collapsed part was easier than access to the collapsed part [39]. Higher enzyme dosage resulted in the biggest difference in digestibility (DEI) at the earlier stage (29.3 at 6 h) because there was a considerable difference in the amount of collapsed pores in the ND and OD pulp. The ND pulp was hydrolyzed quickly due to the small amount of collapsed pores and the fast hydrolysis of the non-collapsed pores in the ND pulp by more enzymes resulted in a big difference in the DEI in 6 h. As the hydrolysis progressed, the reversible collapsed pores were converted to partial collapsed pores by rewetting in buffer solution and the cellulose conversion to sugars induced by progress in the hydrolysis provided more pore volume in the cellulose [40]. Thus, the DEI with 20 FPU decreased as hydrolysis progressed, which meant that there was less difference in enzymatic digestibility between the ND and OD pulp. Therefore, it was concluded that more enzymes caused a higher drying effect at the initial stage of enzymatic hydrolysis and the time for the highest drying effect as revealed earlier. Once the drying effect reached maximum, the drying effect was reduced as hydrolysis progressed when 10 and 20 FPU of enzymes were used. When 4 FPU of enzymes was used, however, there was little difference in the DEI as hydrolysis progressed, unlike 10 and 20 FPU of enzymes. It was considered that the slower rate of hydrolysis with less enzymes caused more inhibition due to a decrease in enzyme activity, and thus, new pores formed during hydrolysis could not dramatically enhance enzyme accessibility in 120 h. In summary, the drying effect on enzymatic hydrolysis increased as the enzyme dosage increased, and the highest DEI occurred in the earlier stages with higher enzyme dosage. As enzymatic hydrolysis with 10 and 20 FPU of enzymes progressed, the drying effect decreased gradually due to the formation of pores in the OD pulp during hydrolysis. When 4 FPU was used, however, a decrease in the DEI was not observed in 120 h because there was not enough pore formation from enzyme inhibition.

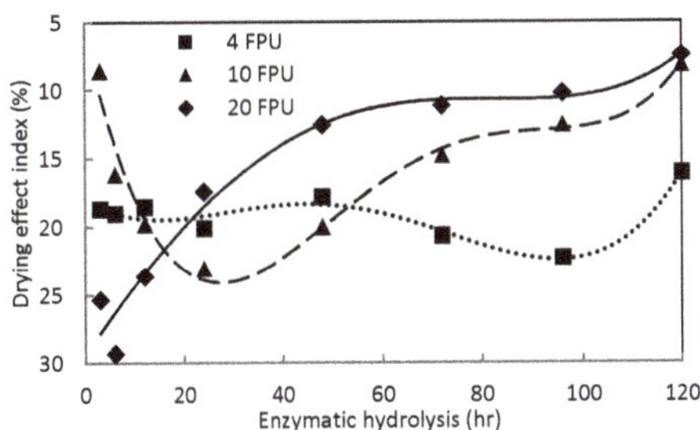

Figure 2. Drying effect index depending on enzyme dosage and hydrolysis time.

3.3. Evaluation of Drying Effect Using Water Retention Value and Simons' Staining Method

The water retention value (WRV) has been used to characterize the drying effect on enzymatic hydrolysis induced by fiber hornification because it can easily measure the pore volume of pulp in a wet state by measuring the amount of water in cell wall [22].

It was previously found that the WRV of the OD pulp decreased significantly due to collapse and shrinkage of pores and the decrease in the WRV correlated well with enzymatic digestibility [22,32]. In this study, the OD pulp with lower enzymatic digestibility, as shown above, revealed a significantly lower WRV, which decreased by 36.4% and 33.3% based on the ND and FD pulps, respectively (Table 3). The decrease in WRV related to the lower enzymatic digestibility was also caused by less enzyme accessibility to cellulose, induced by the decrease in pore volume [22]. However, as enzymatic hydrolysis progresses, new pores are formed through cellulose conversion to sugars and the pore formation increases the pore volume in cellulose. Therefore, the WRV of the OD pulp hydrolyzed for 24 h increased drastically from 105.7 to 203.0%. As described, the kinetics of enzymatic hydrolysis showed that the bigger drying effect at initial stages, which is due to less pore volume caused by the collapse of pores and pores formed by cellulose conversion during hydrolysis, can supply enzymes with space to access cellulose. Therefore, the difference in enzymatic digestibility between the ND and OD pulps will decrease.

Table 3. Water retention value (WRV) and enzymatic digestibility depending on drying.

FPU/g-Pulp	Never Dried	Freeze Dried	Oven Dried	Oven Dried after 24 h of EH with 10 FPU/g-Pulp
WRV (%)	166.1	158.4	105.7	203.0
EH with 4 FPU/g-pulp	83.9 ± 2.2	81.4 ± 0.6	70.4 ± 0.8	NA
EH with 10 FPU/g-pulp	96.6 ± 0.5	96.1 ± 0.4	88.7 ± 0.1	NA
EH with 20 FPU/g-pulp	100.7 ± 0.9	101.0 ± 0.2	93.2 ± 0.4	NA

The Simons' staining method can be used to evaluate the enzyme accessible surface area on cellulose using adsorption and desorption of dyes [26,30]. The pores are segregated depending on the diameter of two different dyes, which have a different affinity for hydroxyl groups in cellulose, and thereby, enzyme accessible surface area of cellulose can be measured in a wet state. The large orange-colored dye molecules can only penetrate larger pores and adsorb in preference to the small blue-colored dye molecules as a result of their greater affinity for hydroxyl groups in cellulose [29]. The large orange-colored dye molecule is mainly composed of two fractions that have hydrodynamic diameters of 5–7 and 12–36 nm [29]. The 5–7 nm diameter is very similar to that of the catalytic core of *Trichoderma reesei* endoglucanase [21]. Therefore, enzyme accessible surface area can be evaluated by the amount of adsorbed orange dye [26], even though the adsorption behavior onto the cellulose surface may be different to enzymes, and also the drying effect on enzymatic hydrolysis has been assessed effectively by the Simons' staining [21].

As result of the Simons' staining, we found that the amount of orange dye adsorbed in the ND and OD pulps were 154.4 and 75.9 mg per g of cellulose, respectively, and the drying process decreased more than half of the accessible surface area in larger pores by pore collapse (Table 4). However, the amount of blue dye, which adsorbed on the surface area of smaller pores, increased from 36.7 to 62.1 mg per g of cellulose. The increase in the adsorption of blue dye was likely caused by the change of larger pores to smaller pores due to the collapse of larger pores, and thus, the surface area of smaller pore increased [21]. The increase in enzyme accessible surface area enhanced enzymatic digestibility through better enzyme adsorption, and it was more remarkable at the initial stage. The total amount of adsorbed dye decreased from 191.0 to 138.0 by drying, which suggested that the formation of the smaller pores and disappearance of larger pores by pore collapse in cellulose decreased the total surface area. It is well known that most pretreatment processes open up the cell wall structure by mechanical shearing force and/or chemical removal of components and this forms new pores in cellulose [41]. Thus, enzymatic hydrolysis could be enhanced by increasing the accessibility through pore formation. The drying process decreased enzyme accessibility and reduced enzymatic hydrolysis. This can be explained by using the opposing mechanism of pretreatment on enzyme accessibility and enzymatic hydrolysis. The drying effect was more remarkable at the initial stage and then it decreased as hydrolysis progressed. However, the irreversible change in structure made a difference to the maximum enzymatic digestibility.

Table 4. Amount of adsorbed dye depending on drying methods.

Pulps	Amount of Adsorbed Dye by Simons' Staining (mg·g^{-1})		
	Orange Dye	Blue Dye	Total
Never dried	154.4 ± 2.3	36.7 ± 2.0	191.0 ± 4.3
Oven dried	75.9 ± 3.0	62.1 ± 5.8	138.0 ± 8.8

3.4. Evaluation of Drying Effect on Crystalline Structure Using ^{13}C CPMAS Solid-State NMR Analysis

The ultrastructure of cellulose has mainly been determined by ^{13}C CPMAS solid-state NMR spectroscopy [42,43], and the solid-state NMR characterized changes associated with drying pulp [44] and bleached pulp by hornification has also been reported [19].

The ^{13}C solid-state NMR analysis was performed in wet conditions for the ND, OD and FD pulps and the spectra with signals assigned are shown in Figure 3 [19]. The ^{13}C NMR spectrum of the pulp was mainly dominated by signals assigned to cellulose. Signals assigned to C-4 and C-6 were partially separated into two clusters, labeled i and s, which were assigned to the interior and surface of the crystalline domains, respectively. Broadening of the NMR signal and relative sharpening of C-4 and C-6 signals on the surfaces of crystalline domains by partial drying has been reported, but the signal broadening and sharpening of C-4 and C-6 signals can be reversed by rewetting the pulp [19]. The ^{13}C NMR spectra of all three pulps were quite similar because both the FD and OD pulps were analyzed in a wet state. Although it has been reported that several repetitions of drying and rewetting cause the difference to peak at 89.4 ppm due to the irreversible drying effect [19], it was not observed in this study because drying was performed only once and rewetting for NMR analysis might reverse the drying effect. Therefore, it can be concluded that drying once did not cause an irreversible change in crystalline structure, and therefore, the decrease in enzymatic digestibility does not seem to be related to the structure of cellulose. However, it might be related to cellulose structure if the pulp was dried more than once.

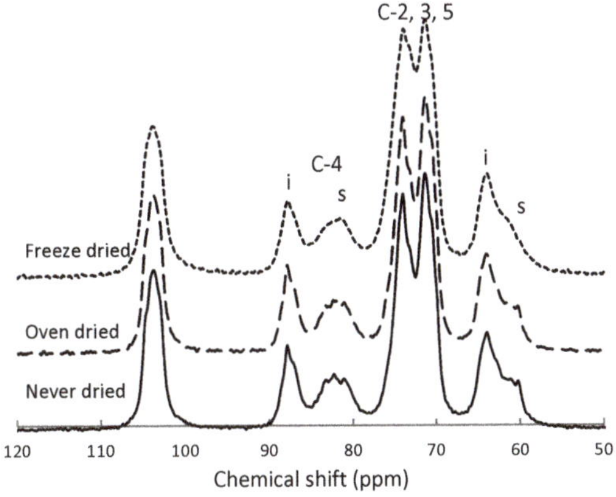

Figure 3. 13C solid-state NMR spectra of never-dried, freeze-dried and oven-dried pulps.

4. Conclusions

The drying effect was evaluated by the difference in enzymatic digestibility in the ND and OD pulp, which was found to vary depending on enzyme dosage. The drying effect index (DEI) was defined as the difference in enzymatic digestibility between the OD and ND pulps and it was concluded that more enzymes cause a higher DEI at the initial stage of enzymatic hydrolysis, and the highest DEI was also observed in the earlier stages with the higher enzyme dosage. Once the DEI reached maximum, it began to reduce as enzymatic hydrolysis progressed due to pore formation by cellulose conversion to sugars. However, there was no big difference in the DEI with lower enzyme dosage because slower hydrolysis with less enzymes caused more enzyme inhibition. Thus, the pores formed by cellulose conversion during hydrolysis with less enzymes could not improve enzyme accessibility after 120 h.

It was also found that the drying effect was well correlated with porosity in cellulose, as measured by the water retention value (WRV) and Simons' staining. The porosity in cellulose was reduced during drying by pore collapse and thus, enzyme accessibility decreased accordingly. The pore collapse was partially reversed by rewetting the pulp, however, there was an irreversible structural change that decreased the final enzymatic digestibility of the OD pulp. However, irreversible change in crystalline structure by drying once was not observed through the 13C solid-state NMR, and thus, decreases in enzymatic digestibility did not seem to be related to the crystalline structure of cellulose.

Author Contributions: Conceptualization, B.K.; methodology, B.K.; formal analysis, B.K. and J.J.; investigation, B.K. and S.-M.C.; resources, B.K.; data curation, S.-M.C.; writing—original draft preparation, B.K.; writing—review and editing, B.K.; visualization, J.J.; supervision, B.K.; project administration, B.K. All authors have read and agreed to the published version of the manuscript.

Funding: This study was carried out with the support of the R & D Program for Forest Science Technology (Project No. 2020226A00-2022-AC01) provided by the Korea Forest Service (Korea Forestry Promotion Institute).

Acknowledgments: The NMR analysis in this work was performed by Rui Katahira, a research chemist at the National Renewable Energy Laboratory.

Conflicts of Interest: The authors declare no conflict of interest.

References

1. Li, T.; Takkellapati, S. The current and emerging sources of technical lignins and their applications. *Biofuel Bioprod. Biorefin.* **2018**, *12*, 756–787. [CrossRef] [PubMed]

2. Bozell, J.J.; Petersen, G.R. Technology development for the production of biobased products from biorefinery carbohydrates-the US Department of Energys "Top 10" revisited. *Green Chem.* **2010**, *12*, 539–554. [CrossRef]

3. Biddy, M.J.; Scarlata, C.; Kinchin, C. *Chemicals from Biomass: A Market Assessment of Bioproducts with Near-Term Potential*; National Renewable Energy Laboratory and Department of Energy: Washington, DC, USA. Available online: https://www.osti.gov/biblio/1244312 (accessed on 10 August 2020). [CrossRef]

4. Wheeldon, I.; Christopher, P.; Blanch, H. Integration of heterogeneous and biochemical catalysis for production of fuels and chemicals from biomass. *Curr. Opin. Biotechnol.* **2017**, *45*, 127–135. [CrossRef] [PubMed]

5. Koo, B.; Treasure, T.H.; Jameel, H.; Phillips, R.B.; Chang, H.M.; Park, S. Reduction of enzyme dosage by oxygen delignification and mechanical refining for enzymatic hydrolysis of green liquor-pretreated hardwood. Appl. *Biochem. Biotechnol.* **2011**, *165*, 832–844. [CrossRef] [PubMed]

6. Mata, T.M.; Martins, A.A.; Caetano, N.S. Bio-refinery approach for spent coffee grounds valorization. *Bioresour. Technol.* **2016**, *247*, 1077–1084. [CrossRef]

7. Nguyen, Q.A.; Cho, E.J.; Lee, D.S.; Bae, H.J. Development of an advanced integrative process to create valuable biosugars including manno-oligosaccharides and mannose from spent coffee grounds. *Bioresour. Technol.* **2018**, *272*, 209–216. [CrossRef]

8. Galbe, M.; Zacchi, G. Pretreatment: The key to efficient utilization of lignocellulosic materials. *Biomass Bioenerg.* **2012**, *46*, 70–78. [CrossRef]

9. Rajendran, K.; Taherzadeh, M.J. Chapter 3 Bioprocessing of Renewable Resources to Commodity Bioproducts. In *Pretreatment of Ligocellulosic Materials*; Wiley: Hoboken, NJ, USA, 2014.

10. Mandels, M.; Hontz, L.; Nystrom, J. Enzymatic hydrolysis of waste cellulose. *Biotechnol. Bioeng.* **1974**, *16*, 1471–1493. [CrossRef]

11. Fernandes Diniz, J.M.B.; Gil, M.; Castro, J. Hornification-its origin and interpretation in wood pulps. *Wood Sci. Technol.* **2004**, *37*, 489–494. [CrossRef]

12. Park, S.; Venditti, R.A.; Jameel, H.; Pawlak, J.J. A novel method to evaluate fibre hornification by high resolution thermogravimetric analysis. *Appita J.* **2006**, *59*, 481–485.

13. Ferreira, S.R.; Silva, F.; Lima, P.R.L.; Filho, R.D.T. Effect of hornification on the structure, tensile behavior and fiber matrix bond of sisal, jute and curauá fiber cement based composite system. *Constr. Build. Mater.* **2017**, *139*, 551–561. [CrossRef]

14. Kato, K.; Cameron, R. A review of the relationship between thermally-accelerated ageing of paper and hornification. *Cellulose* **1999**, *6*, 23–40. [CrossRef]

15. Weise, U. Hornification: Mechanisms and terminology. *Paperi Ja Puu.* **1998**, *80*, 110–115.

16. Welf, E.; Venditti, R.; Hubbe, M.; Pawlak, J. The effects of heating without water removal and drying on the swelling as measured by water retention value and degradation as measured by intrinsic viscosity of cellulose papermaking fibers. *Prog. Pap. Recycl.* **2005**, *14*, 1–9.

17. Chen, Y.; Wan, J.; Ma, Y. Effect of noncellulosic constituents on physical properties and pore structure of recycled fibre. *Appita J.* **2009**, *62*, 290–295.

18. Nazhad, M.M. Fundamentals of Strength Loss in Recycled Paper. Ph.D. Thesis, University of British Columbia, Vancouver, BC, Canada, 22 December 1994. V6T 1Z4. [CrossRef]

19. Newman, R.H. Carbon-13 NMR evidence for cocrystallization of cellulose as a mechanism for hornification of bleached kraft pulp. *Cellulose* **2004**, *11*, 45–52. [CrossRef]

20. Wistara, N.; Young, R.A. Properties and treatments of pulps from recycled paper. Part, I. Physical and chemical properties of pulps. *Cellulose* **1999**, *6*, 291–324. [CrossRef]

21. Esteghlalian, A.; Bilodeau, M.; Mansfield, S.; Saddler, J. Do enzymatic hydrolyzability and Simons' stain reflect the changes in the accessibility of lignocellulosic substrates to cellulase enzymes? *Biotechnol. Prog.* **2001**, *17*, 1049–1054. [CrossRef]

22. Luo, X.; Zhu, J. Effects of drying-induced fiber hornification on enzymatic saccharification of lignocelluloses. *Enzyme Microb. Technol.* **2011**, *48*, 92–99. [CrossRef]

23. Li, Y.; Li, B.; Mo, W.; Yang, W.; Wu, S. Influence of residual lignin and thermal drying on the ultrastructure of chemical hardwood pulp and its enzymatic hydrolysis properties. *Cellulose* **2019**, *26*, 2075–2085. [CrossRef]

24. TAPPI T205 sp-95. *Forming Handsheets for Physical Tests of Pulp*; TAPPI Press: Atlanta, GA, USA, 1995; Available online: https://www.tappi.org/content/sarg/t205.pdf (accessed on 10 August 2020).

25. NREL. In *Determination of Structural Carbohydrates and Lignin in Biomass*; Laboratory Analytical Procedure (LAP), US Department of Energy, National Renewable Energy Laboratory: Golden, CO, USA, 2008. Available online: https://www.nrel.gov/docs/gen/fy13/42618.pdf (accessed on 10 August 2020).

26. Koo, B.; Jameel, H.; Phillips, R.; Chang, H.; Park, S. Reduction of enzyme dosage for enzymatic hydrolysis by the mechanical refining and oxygen bleaching post-treatments. *Appl. Biochem. Biotechnol.* **2011**, *165*, 832–844. [CrossRef] [PubMed]

27. Koo, B.; Park, S. A method to evaluate biomass accessibility in wet state based on thermoporometry. *Methods Mol. Biol.* **2012**, *908*, 83–89.

28. TAPPI UM256. *Water Retention Value (WRV)*; TAPPI Press: Atlanta, GA, USA, 1981; Available online: https://imisrise.tappi.org/TAPPI/Products/01/UM/0104UM256.aspx (accessed on 10 August 2020).

29. Yu, X.; Atalla, R. A staining technique for evaluating the pore structure variations of microcrystalline cellulose powders. *Powder Technol.* **1998**, *98*, 135–138. [CrossRef]

30. Chandra, R.; Ewanick, S.; Hsieh, C.; Saddler, J. The characterization of pretreated lignocellulosic substrates prior to enzymatic hydrolysis, part 1: A modified Simons' staining technique. *Biotechnol. Prog.* **2008**, *24*, 1178–1185. [CrossRef] [PubMed]

31. Lee, Y.H.; Fan, L. Kinetic studies of enzymatic hydrolysis of insoluble cellulose: Analysis of the initial rates. *Biotechnol. Bioeng.* **1982**, *24*, 2383–2406. [CrossRef]

32. Ioelovich, M.; Morag, E. Effect of cellulose structure on enzymatic hydrolysis. *BioResources* **2011**, *6*, 2818–2835.

33. Sun, S.; Sun, S.; Cao, X.; Sun, R. The role of pretreatment in improving the enzymatic hydrolysis of lignocellulosic materials. *Bioresour. Technol.* **2016**, *199*, 49–58. [CrossRef]

Appl. Sci. **2020**, *10*, 5545

34. Hribernik, S.; Kleinschek, K.S.; Rihm, R.; Ganster, J.; Fink, H.P.; Smole, M.S. Tuning of cellulose fibres' structure and surface topography: Influence of swelling and various drying procedures. *Carbohydr. Polym.* **2016**, *148*, 227–235. [CrossRef]
35. Buschle-Diller, G.; Fanter, C.; Loth, F. Structural changes in hemp fibers as a result of enzymatic hydrolysis with mixed enzyme systems. *Text. Res. J.* **1999**, *69*, 244–251. [CrossRef]
36. Kristensen, J.B.; Felby, C.; Jørgensen, H. Yield-determining factors in high-solids enzymatic hydrolysis of lignocellulose. *Biotechnol. Biofuels* **2009**, *2*, 11. [CrossRef]
37. Jing, X.; Zhang, X.; Bao, J. Inhibition performance of lignocellulose degradation products on industrial cellulase enzymes during cellulose hydrolysis. *Appl. Biochem. Biotechnol.* **2009**, *159*, 696–707. [CrossRef] [PubMed]
38. Spinu, M.; Dos Santos, N.; Le Moigne, N.; Navard, P. How does the never-dried state influence the swelling and dissolution of cellulose fibres in aqueous solvent? *Cellulose* **2011**, *18*, 247–256. [CrossRef]
39. Wang, Q.; He, Z.; Zhu, Z.; Zhang, Y.H.P.; Ni, Y.; Luo, X.; Zhu, J. Evaluations of cellulose accessibilities of lignocelluloses by solute exclusion and protein adsorption techniques. *Biotechnol. Bioeng.* **2011**, *109*, 381–389. [CrossRef] [PubMed]
40. Zeng, M.; Mosier, N.S.; Huang, C.P.; Sherman, D.M.; Ladisch, M.R. Microscopic examination of changes of plant cell structure in corn stover due to hot water pretreatment and enzymatic hydrolysis. *Biotechnol. Bioeng.* **2007**, *97*, 265–278. [CrossRef] [PubMed]
41. Lee, S.; Teramoto, Y.; Endo, T. Enzymatic saccharification of woody biomass micro/nanofibrillated by continuous extrusion process I-Effect of additives with cellulose affinity. *Bioresour. Technol.* **2009**, *100*, 275–279. [CrossRef]
42. Atalla, R.; VanderHart, D. The role of solid state 13C NMR spectroscopy in studies of the nature of native celluloses. *Solid State Nucl. Magn. Reson.* **1999**, *15*, 1–19. [CrossRef]
43. Hult, E.L.; Iversen, T.; Sugiyama, J. Characterization of the supermolecular structure of cellulose in wood pulp fibres. *Cellulose* **2003**, *10*, 103–110. [CrossRef]
44. Hult, E.L.; Larsson, P.; Iversen, T. Cellulose fibril aggregation—an inherent property of kraft pulps. *Polymer* **2001**, *42*, 3309–3314. [CrossRef]

Review

Lignin to Materials: A Focused Review on Recent Novel Lignin Applications

Osbert Yu [1] and Kwang Ho Kim [2,3,*] 

[1]   Department of Chemical and Biological Engineering, University of British Columbia, Vancouver, BC V6T 1Z4, Canada; osbert.yu@alumni.ubc.ca
[2]   Department of Wood Science, University of British Columbia, Vancouver, BC V6T 1Z4, Canada
[3]   Clean Energy Research Center, Korea Institute of Science and Technology, Seoul 02792, Korea
*   Correspondence: kwanghokim@kist.re.kr

Received: 4 June 2020; Accepted: 2 July 2020; Published: 3 July 2020

Abstract: In recent decades, advancements in lignin application include the synthesis of polymers, dyes, adhesives and fertilizers. There has recently been a shift from perceiving lignin as a waste product to viewing lignin as a potential raw material for valuable products. More recently, considerable attention has been placed in sectors, like the medical, electrochemical, and polymer sectors, where lignin can be significantly valorized. Despite some technical challenges in lignin recovery and depolymerization, lignin is viewed as a promising material due to it being biocompatible, cheap, and abundant in nature. In the medical sector, lignins can be used as wound dressings, pharmaceuticals, and drug delivery materials. They can also be used for electrochemical energy materials and 3D printing lignin–plastic composite materials. This review covers the recent research progress in lignin valorization, specifically focusing on medical, electrochemical, and 3D printing applications. The technoeconomic assessment of lignin application is also discussed.

Keywords: lignin valorization; lignin applications; 3D printing; electrochemical material; medical application

---

1. Introduction

The overdependence on fossil fuels has raised increasing concerns about climate change and an energy crisis, which has warranted research to search for renewable and clean energy alternatives. Lignocellulosic biomass is the most abundant and renewable source of organic carbon on Earth, presenting the best option to achieve a sustainable biorefinery in the future [1]. In recent decades, there has been significant research conducted to convert biomass components into biofuel and value-added products. Although some sectors in biorefineries, including lignocellulosic bioethanol, have been widely researched, the economic feasibility is often discussed when it comes to commercialization. From the technoeconomic perspective, the success of future biorefineries is highly dependent on lignin valorization.

Lignin can be obtained from a variety of natural sources, including woody biomass, agricultural residues, and energy crops. Regardless of the type of lignin, there are typically two pathways for lignin valorization. One pathway uses the lignin as a macro-polymer to produce valuable materials; the other pathway involves the depolymerization of lignin into low-molecular weight monomers [2]. Lignin monomers can undergo derivatization through various chemical processes to be converted into desired products [3].

The world annually produces around 100 million metric tons of lignin, worth approximately 732.7 million USD [3]. Since lignin is a cheap and abundant natural polymer, much research has been spent on valorizing lignin. In recent decades, perspectives on lignin have changed from a waste product used as a low-grade fuel and animal feed to valuable products such as polymers, adhesives, and others [3–6]. More recently, lignin valorization has received interest in other sectors, like the medical and electrochemical energy materials sectors. The scope of this review will focus on the valorization of technical lignin in advanced materials.

2. Technical Lignins

In nature, lignin is an amorphous phenolic polymer that is randomly branched and crosslinked with cellulose and hemicellulose [7]. To utilize lignin, it needs to be isolated from biomass through various processes, and the extracted lignin is called technical lignin. Since most technical lignins are available in the pulp and paper industries as byproducts, the production of lignin will continue to increase as lignocellulosic ethanol emerges. Considering that the lignin extraction method modifies the native structure of lignin, it is important to understand how the process affects lignin structures to develop effective lignin valorization technologies. There have been many review articles that thoroughly discuss lignin separation methods and their influence on the structures and properties of the extracted lignin [8–11]. The following overview briefly introduces common industrial processes in which lignin is extracted from biomass.

Figure 1 illustrates several processes to extract technical lignins from lignocellulosic biomass. Among the technical processes, Kraft pulping is the major chemical pulping process, accounting for 85% of the total lignin production in the world [12]. The process is performed at a high pH, and about 90–95% of the lignin is dissolved into the black liquor. Kraft lignin is typically precipitated and recovered from black liquor by the addition of acidifying agents. Predominantly, the acidification is carried out by adding either mineral acid (e.g., sulfuric acid) or carbon dioxide, followed by filtering, washing, and drying for the recovery of Kraft lignin. Around 630,000 tons of Kraft lignin are annually produced, and most Kraft lignin is combusted for heat generation, resulting in low-value utilization. The sulfite pulping process is conducted between a pH of 2–12, depending on the cationic composition of the pulping liquor [13]. Lignosulfonates, isolated lignins from the sulfite process, contain significant amounts of sulfur in the form of sulfonate groups. Since lignosulfonates are widely available, lignosulfonates were used in a wide range of applications, such as dispersants, flocculants, concrete additives, and composites [14].

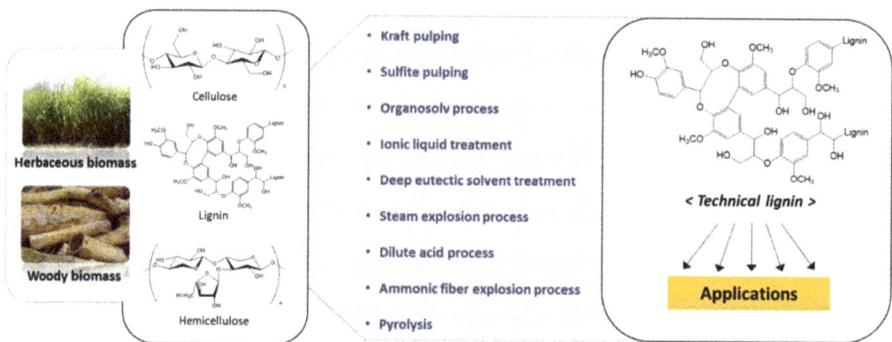

Figure 1. Processes for the extraction of technical lignins.

The organosolv process constitutes the fractionation of biomass components through treatment using an organic solvent, such as ethanol, ethylene glycol, acetone, tetrahydrofuran, and γ-valerolactone [15,16]. Since the organosolv process is conducted in the absence of sulfur, it has recently been utilized more so than Kraft and sulfite pulping. Furthermore, the large-scale production of organosolv lignin is expected from the emerging cellulosic ethanol sectors, which offers significant opportunities for lignin valorization.

In addition to the conventional pulping and biomass pretreatment processes such as steam explosion, dilute acid, and ammonic fiber explosion, there have been significant advances in biomass fractionation to extract high-quality lignin using novel solvents. For example, ionic liquids (ILs) and deep eutectic solvents (DESs) have attracted considerable attention as promising agents for biomass fractionation due to their high solvation capacity in the dissolution of biomass components [17,18]. The technical lignins obtained from ILs and DESs have been found to retain their original structures; the eco-friendly properties of such solvents offer new options for lignin extraction.

3. Medical Applications

3.1. Wound Dressings

Hydrogels are three-dimensional (3D) networks of polymers; their hydrophilic structures make them capable of absorbing and holding large amounts of water in their 3D networks [19,20]. Hydrogels have gained significant attention in biomedical sectors because they can be used for drug delivery, tissue engineering, and antimicrobial materials [21]. Recently, lignin has been viewed as a promising material for the production of hydrogels, because lignin possesses great antioxidant and antibacterial properties [22]. Furthermore, hydrogels have a high absorption capacity, allowing them to effectively remove undesirable metabolites from the wound [23]. Due to lignin's high mechanical strength, the use of lignin in hydrogels helps protect the wound from further injury or contamination [23]. However, due to the relatively low amount of hydroxyl groups and its rigid structure, lignin requires pretreatment for it to be effectively used as a medical material. In this respect, many studies have focused on developing chemical modification strategies of lignin. Such modifications include the introduction of new active sites on lignin by combining other materials and the functionalization of the hydroxyl group to enhance its reactivity.

There are several ways that lignin can be combined with other materials to construct composite wound dressings that are antimicrobial and biocompatible. A study that used a lignin model polymeric compound, dehydrogenate polymer (DHP), in alginate (Alg) hydrogel, determined that the lignin model DHP in hydrogel produced antimicrobial effects against several clinical bacterial strains and did not have toxic effects on human epithelial cells [24]. A stock suspension of 10 mg/mL DHP and 20 mg/mL Alg was used to test for antibacterial properties. The DHP–Alg demonstrated minimum inhibitory concentration values of 0.002–0.90 mg/mL and a minimum bactericidal concentration of 0.004–1.25 mg/mL. Furthermore, DHP–Alg showed higher antibacterial activity against *L. monocytogenes*, *P. aeruginosa*, and *S. Typhimurium* than streptomycin and ampicillin [24]. The mechanism of antibacterial action was speculated to be that DHP in hydrogel interacts with bacterial cell wall synthesis and/or cell wall structure, which could lead to the disorganization of the cell wall. Although more studies need to be conducted on elucidating the antibacterial mechanism of DHP, the results show the potential application of lignin as a wound-healing agent.

Lignin amine and sulfite-pulped lignin can also be used to construct two different wound dressings. Lignin amine can be synthesized from sodium lignosulfonate by a Mannich reaction; the lignin amine can be crosslinked with poly(vinyl alcohol) (PVA) to form a hydrogel. When a solution of silver nitrate was added to the lignin-based hydrogel, the biocompatible hydrogel not only demonstrated enhanced antimicrobial properties against *E. coli* and *S. aureus*, but the hydrogel also exhibited good elasticity [25]. Additionally, the lignin obtained from sulfite pulping can be combined with chitosan dissolved in an acetic acid solution and PVA dissolved in water to form a lignin–chitosan–PVA composite hydrogel. The sulfonate groups in the lignin formed ionic bonds with amino groups in the chitosan; this is what gives the hydrogel a high mechanical strength and high antioxidant activity [23]. The lignin–chitosan–PVA composite hydrogel demonstrated inhibitory effects on *S. aureus* by penetrating its cell membrane. Results from the study suggested that a hydrogel with a greater lignin concentration resulted in an increase in the swelling ratio, hydrophilicity, protein adsorption capacity, tensile strength, and elongation.

Three-dimensional-printed wound dressings can also be constructed from lignin. Poly(lactic acid) (PLA) pellets coated with castor oil can also be combined with a small amount of Kraft lignin (3 wt % or less) and an antibiotic, tetracycline, to create a composite used for 3D printing wound dressings. The printed filament material was treated with the 2,2-diphenyl-1-picrylhydrazyl (DPPH), a free radical standard, to assess its radical scavenging activity, which can potentially be applied in wound care applications [26]. Figure 2 shows photographs of meshes that were 3D-printed from PLA and 2 wt % Kraft lignin. While lignin in the 3D-printing composite provided antioxidant properties to the composite, small amounts of lignin in the composite did not demonstrate antimicrobial properties with *S. aureus*. However, the tetracycline in the composite was what contributed to the composite's antimicrobial properties. More interestingly, the mesh of the PLA–tetracycline–lignin wound dressing can be 3D-printed based on the patients' needs, leading to the possible customization of wound dressings in the future.

Figure 2. Meshes 3D-printed from poly(lactic acid) (PLA) and 2 wt % lignin with grid sizes of 1.5 mm and 1 mm. The middle column shows the side view of each mesh when slightly bent, while the right column shows the close-ups of each mesh. Reprinted from reference [26], with permission from the MDPI.

3.2. Pharmaceuticals

There are specific compounds derived from lignin that can not only be used to construct pharmaceuticals to alleviative the symptoms of diseases, but can also be used to construct drug delivery materials. While sulfite and Kraft lignins are more commonly used lignins to synthesize pharmaceutical products, recently lignins obtained from steam explosion and the organosolv process have been investigated for pharmaceutical use. The mechanism commonly employed to utilize lignin is to depolymerize lignin and obtain derivatives to construct biologically active compounds. These biologically active compounds can be used to treat the Herpes simplex virus, influenza virus, and other viruses [27]. Since the toxicity of drugs is an important factor in the drug discovery phase, the cytotoxicity of lignin was studied; findings from a study show that lignins are generally safe to consume and no not disrupt cell viability [27]. When synthesizing pharmaceutical drugs, the antioxidant property of lignins is highly valued because its hydroxyl functional groups in their phenolic rings neutralize free radicals and protect molecules from oxidation [27]. For instance, lignins isolated from Bagasse, steam explosion lignins, and lignosulfonates have half maximal inhibitory concentrations of 44.9 µg/mL, 74.6 µg/mL, and 133.6 µg/mL, respectively. Compared to a common antioxidant present in tea, epicatechin, which has a half maximal inhibitory concentration of 42.3 µg/mL, these lignins have enhanced antioxidant properties [28].

Lignin-derived components and products, including polyphenols, may help in controlling diseases, like diabetes, coronary heart disease, and Alzheimer's disease. Results of a study suggest that lignosulfonic acid can be taken with glucose to cause a delay in the uptake of glucose to potentially treat diabetes [29]. The study administered a combination of lignosulfonic acid and glucose as well as glucose alone to determine their effects on 2-deoxyglucose uptake in human colorectal adenocarcinoma cells. Since lignosulfonic acid is a non-competitive inhibitor of α-glucosidase, lignosulfonic acid can inhibit α-glucosidase activity and delay the absorption of glucose in the intestines. Lignosulfonic acid was also demonstrated to inhibit human immunodeficiency virus (HIV) and herpes simplex virus (HSV) transmission when tested with human T-cell leukemia cells, human embryonic kidney cells, and peripheral blood mononuclear cells [30]. Furthermore, lignosulfonates have the potential to be developed into drugs that could decrease oxidative activity, thereby boosting the immune system [3]. Lignin may also helpful in controlling obesity. Out of the dietary fiber components, lignin is reported to be the strongest bile acid adsorbent due to the presence of methoxyl and β-carbonyl groups in lignin. When the lignin binds to the bile, micelles are unable to be produced, thereby resulting in decreased lipid absorption [27].

Table 1 summarizes several studies describing the value products that can be synthesized from technical lignins. In a study, lignophenols derived from the native lignin of Japanese cedar were shown to decrease oleate-induced apolipoprotein-B secretion and reduce cholesterol in HepG2 cells (human liver cells), thereby potentially preventing coronary heart disease [31]. These lignophenols were derived using a phase separation process that included cresol and sulfuric acid; lignophenols were then dissolved in dimethyl sulfoxide before being cultured with HepG2 cells. Another study presents three steps in which lignocellulose can be depolymerized to form platform chemicals, which can then be used to synthesize biologically active compounds with the use of a DES [32]. The DES used was a mixture of choline chloride and oxalic acid. Biologically active compounds that can be synthesized include tetrahydro-2-benzazepines. Tetrahydro-2-benzazepines are present in alkaloids, such as galantamine, that could treat Alzheimer's disease. In this three-step process, the only by-product was water. Furthermore, biologically active compounds derived in this study were also shown to be effective against *S. aureus*. Figure 3 summarizes the reaction pathway to synthesize biologically active compounds that likely exhibit antibacterial or anticancer activities.

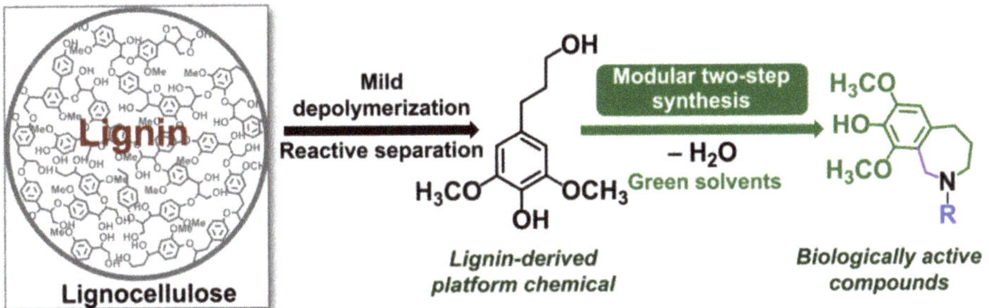

Figure 3. Summary of the reaction pathway to synthesize the biologically active compounds. Reprinted with permission from [32]. Copyright (2019) American Chemical Society.

Table 1. Summary of the studies on the pharmaceutical applications of lignin.

Product	Usage	Feedstock Lignin	Lignin Content (wt %)	Reaction/Process Condition	Source
Lignophenols	Potentially prevent coronary heart disease	Japanese cedar native lignin	N/A	Phase-separation with cresol and sulfuric acid.	[31]
Tetrahydro-2-benzazepines	Precursor to the drugs that could treat Alzheimer's disease	Pine and poplar lignocellulose	N/A	For depolymerization step, copper-doped porous metal oxides and 40 bar H_2; for synthesis of product, used choline chloride/ oxalic acid DES at 70–80 °C for 20–48 h.	[32]
Drug vehicles	Deliver methotrexate	Sugar cane bagasse	N/A	Lignin soaked in 100 ppm methotrexate solution with intense stirring at 40 °C for 18 h.	[33]
Nanoparticles	Deliver Sorafenib and Benzazulene	LignoBoost™ softwood Kraft lignin	50%	Lignin dissolved in tetrahydrofuran and put in the dialysis bag, iron (III) isopropoxide in tetrahydrofuran for 24 h.	[34]
Hydrogels	Drug delivery system	Kraft lignin	Up to 80%	20 to 200 wt % citric acid, used micro-extruder at 120 °C for 2 or 5 min of recirculation.	[35]
Biosorbent	Adsorb ibuprofen and acetaminophen	Kraft lignin	50%	Kraft lignin and α-chitin powder mixed, following activation with 15% hydrogen peroxide.	[36]
Photocatalyst	Degrade acetaminophen	LignoTech lignin	33%	TiO_2 and NaOH, hydrothermal treatment at 130 °C for 48 h, centrifuged, dried, calcined in N_2 or air.	[37]
Nanofibers	Adsorb fluoxetine	Alkali lignin	30–50%	Aqueous polyvinyl alcohol and distilled water at 80 °C for 60 min, electrospinning fibers.	[38]

Materials containing lignin and lignin derivatives can also be effectively used to deliver drugs inside the human body. Perhaps the largest advantages of lignin-based nanoparticles for delivering drugs are their inexpensive and non-toxic properties. For instance, lignins derived from sugarcane can be used as a drug carrier to deliver methotrexate and treat rheumatoid arthritis [33]. Lignin nanoparticles encapsulating iron oxide could effectively deliver Sorafenib and Benzazulene in alkali media [34]. Hydrogels produced from varying concentrations of starch, Kraft lignin, hemicellulose, and citric acid were tested for their ability to deliver drugs effectively [35]. To characterize the performance of these hydrogels, the swelling capability was analyzed because it indicates the crosslinking density and ultimately the amount of water absorbed in the hydrogel. Based on the experiments testing the diffusion of water into the gel, the hydrogel has the potential to swell to 1380% at a pH of 9 and the potential to swell to 345% at a pH of 4. The pH-dependent swelling behavior is desirable for hydrogels, because the diffusion rates of molecules in and out of the gel can be controlled.

Since pharmaceutical products are a common source of water pollution, lignin can adsorb these pharmaceuticals to alleviate water pollution. Kraft lignin and α-chitin powder from crab shells were mixed with hydrogen peroxide to create a sorbent that adsorbed ibuprofen and acetaminophen, which are commonly used drugs that cause water pollution [36]. Figure 4 summarizes the process to construct the biosorbent with a high removal efficiency of ibuprofen and acetaminophen. Several mechanisms for the adsorption process were proposed. One such mechanism was ion–dipole interactions; there could be electrostatic interaction between pharmaceutical ions and the negatively or positively charged surface of the sorbent surface. Another mechanism could be hydrogen bonds forming between the pharmaceutical drug and the sorbent. Such sorbents can potentially be reused. For instance, ethanol was an effective eluent for ibuprofen with a yield of 82.2%, while methanol was an effective eluent for acetaminophen with a yield of 80.8%.

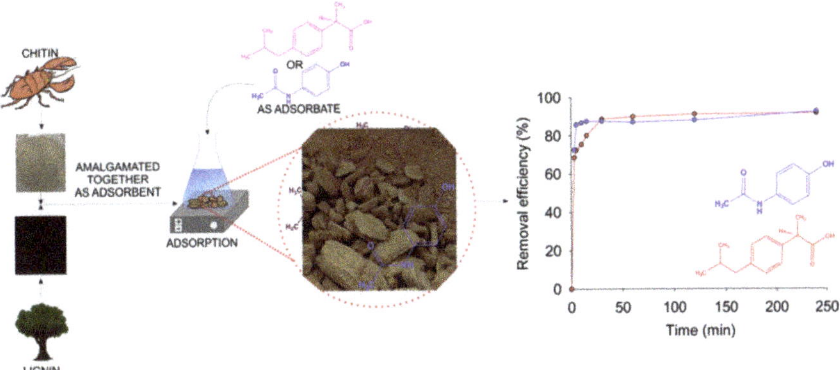

Figure 4. Biosorbent consisting of chitin and lignin removing ibuprofen and acetaminophen with high efficiency. Reprinted from reference [36], with permission from Elsevier.

In another application, lignin was used as a carbonaceous precursor with TiO_2 to form a photocatalyst that degraded acetaminophen, or Tylenol, in one hour of solar radiation [37]. Photocatalysts that experienced thermal treatment in nitrogen rather than in air maintained more carbon from the lignin. This caused a higher light absorption and decreased the photocatalytic efficiency. In another study investigating the use of lignin to adsorb fluoxetine, alkali lignin with low sulfur content and PVA were dissolved to form a solution used for electrospinning [38]. Nanofibers recovered from the electrospinning process then went through a thermostabilization processes and an acidic bath to increase the strength of the fibers. A lignin:PVA ratio of 1:1 resulted in the optimal adsorption capacity; this lignin–PVA nanofibrous membrane adsorbed around 38% of the fluoxetine. The authors of the study hypothesized the sorption mechanism to be the phenol groups in lignin-forming hydrogen bonds with the amino or Fluor groups of the fluoxetine.

4. Electrochemical Energy Materials

In the past decade, lignin has been increasingly investigated for its potential incorporation in the production of battery materials and supercapacitors for the main advantage of being environmentally friendly. With the increasing manufacturing of lignin in recent years, lignin has been available at a low cost. The relatively low cost of lignin makes it an attractive ingredient for the production of anodes for lithium batteries, gel electrolytes, binders, and sodium batteries [39]. Lignin is also valued for the production of energy materials due to its high carbon content that is greater than 60 wt %. Specific functional groups of lignin, such as benzyl and phenolic groups, act as active reaction sites for ions to be stored in applications to supercapacitors. Additionally, the abundant oxygen atoms in lignin are integral for facilitating electrolyte ion adsorption and redox reactions for supercapacitors. Lignin can be used for porous carbon structures in supercapacitors due to its favorable crosslinked structure [2]. While lignin types like lignosulfonate can act as a sulfur-doped agent in batteries or supercapacitors, alkali lignins have shown to be suitable for electrospinning processes to construct nanomaterials [39]. There are generally two processes commonly employed to derive carbon from lignin for energy material applications. One process utilizes precursor carbonization and carbon activation, while the other process uses chemical activators before carbonization and activation that happen at the same time [2]. Graphene, a promising carbon material for electrochemical energy materials, can be synthesized from lignin by catalytic graphitization, carbonization, or oxidative cleavage along with aromatic refusion [40–42].

The incorporation of lignin results in a high gravimetric capacitance and great cycling durability for energy materials. Table 2 shows the summary of recent studies on the electrochemical applications of technical lignins. In an experiment with electrospun carbon nanofibers produced from alkali lignin–PVA solutions, the capacitance of the supercapacitor reduced by 10% after 6000 cycles of discharging and charging [43]. Furthermore, a higher energy value of 42 W h kg^{-1} and a power density of 91 kW kg^{-1} were reported. The experiment suggested that an increasing amount of lignin in precursor nanofibers resulted in a decreased average pore size, increased pore volume, and increased specific surface area. In another study, electrodes constructed from 75 wt % alkali lignin and 25 wt % 0.5 M sodium sulfate electrolyte resulted in one of the highest specific capacitances for electrodes produced from biopolymers: 205 F g^{-1} [44]. The mesopore range of carbon fibers had a wide pore distribution, which contributed to excellent electrochemical performance. Additionally, hierarchical porous carbons can be derived from steam explosion lignin through the carbonization–activation method. Lignin-based hierarchical porous carbons showed a high capacitance of 286.7 F g^{-1} at 0.2 A g^{-1} [45]. The structure of lignin provided advantages to the electrochemical performance, including accessible ion transportation pathways and a high surface area.

Lignin-derived materials used to construct batteries have performed similarly with commercial graphite and other materials commonly used in the industry. Ball-milled hydrolysis lignin can be used for low-rate power sources. A study investigated the performance of a battery that used hydrolysis lignin as the lithium battery cathode material [46]. The cathode material had 76 wt % hydrolysis lignin, 13 wt % carbon black, and 11 wt % polytetrafluoroethylene (PTFE)-based binder. Lithium batteries using this hydrolysis lignin as a cathode material achieved a high discharge capacity of 450 m A h g^{-1}.

Additionally, acetone was used to extract lignin from a corn stalk lignin precursor; the extracted lignin was used to construct hard carbon materials through stabilization in nitrogen, carbonization in nitrogen, and hydrogen reduction [47]. Figure 5 shows the pyrolysis reaction mechanism that occurred after the lignin was extracted from a precursor with acetone. The hard carbon had an initial discharge capacity of 882.2 m A h g^{-1} at 0.1 °C and retained a charge capacity of 228.8 m A h g^{-1} at 2 °C for 200 cycles. In another study, a blend of lignin, PLA, and elastomeric polyurethane was used to create carbon nanofibers through electrospinning. These nanofibers led to increased porosity levels and additional lithium storage sites [48].

Table 2. Summary of the studies on the electrochemical applications of lignin.

Product	Feedstock	Lignin Content (wt %)	Reaction/Process Condition	Source
Supercapacitor	Alkali lignin	30–70%	Electrospinning aqueous alkali lignin and PVA, temperature stabilization in the tube furnace ramped from 25 to 220 °C, held at 220 °C for 8 h.	[43]
Supercapacitor	Softwood alkali lignin	75%	Lignin, PVA, and distilled water mixed at 60 °C for 1 h and at room temperature for 4 h, followed by electrospinning.	[44]
Supercapacitor	Steam explosion lignin	17–50%	Carbonization at 500 °C or 800 °C, activation with KOH, post-treatment by washing with hot water and drying at 100 °C overnight.	[45]
Lithium battery cathode material	Hydrolysis lignin	76%	Ball milled to 30 microns, washed with distilled water in centrifuge for 10–12 h, dried at 60 °C for 24 h.	[46]
Lithium battery anode material	Acetone lignin from corn stalks	80%	Stabilized at 300 °C for 2 h in N_2 in tube furnace.	[47]
Lithium battery anode material	Organosolv hardwood lignin	50–80%	Thermoplastic elastomeric polyurethane, dimethylformamide, PLA stirred at 50 °C for 5 min before electrospinning.	[48]

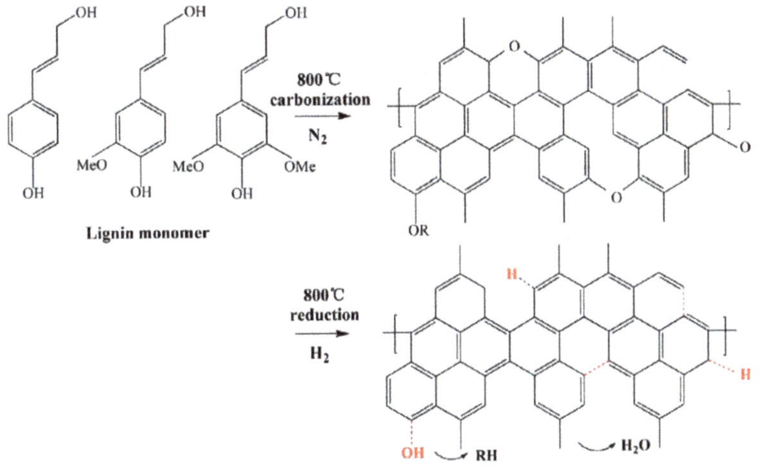

Figure 5. Reaction mechanism using the pyrolysis technology. Reprinted from reference [47], with permission from Elsevier.

5. 3D Printing Lignin–Plastic Composites

Blending lignin with various plastic materials to form 3D printing composites has recently received attention. The ideal 3D printing material has excellent extrudability to ease the process of 3D printing while also being strong so that the final printed material can retain its structure. Lignin has several structures that are advantageous for 3D printing: aliphatic ether groups, β-O-4′ linkages, and oxygenated aromatic bonds [49,50]. Incorporating lignin into plastic materials traditionally used for 3D printing results in composites that can be used for a more environmentally friendly and cost-effective 3D printing process.

There has also been recent discussion with energy sacrificial bonds as a mechanism for strong biomaterials, including lignin. Such sacrificial bonds dissipate energy through rupturing and reform by stretching. In light of this, a study investigated Zn-based coordination bonds between lignin nanoparticles and an elastomer matrix [51]. Such bonds were found to facilitate the dispersion of lignin in the matrix, thereby enhancing the strength of the composite. This study reported that a lignin loading as high as 30 wt % can increase the strength, ductility, and toughness of thermoplastic elastomers, which can be used for 3D printing.

Several studies show that lignin can enhance structural properties of 3D printing materials. Table 3 summarizes recent studies on the application of technical lignins to 3D printing materials. In a study, when Kraft lignin was blended with acrylonitrile butadiene styrene (ABS), the result was a more brittle structure; this composite demonstrated low tensile energy and strength [49]. However, when 10 wt % acrylonitrile butadiene rubber (NBR41) was added to the composite, the chemical and physical crosslinks between the NBR41 and lignin resulted in the improved mechanical properties of the composite relative to common petroleum-based thermoplastics. This composite containing 40 wt % lignin, 10 wt % NBR41, and 50 wt % ABS showed great 3D-printability.

Table 3. Summary of the studies on the 3D printing applications of lignin.

Lignin Feedstock	Polymer Feedstock	Lignin Content in 3D Printing Composite (wt %)	Source
Kraft lignin	ABS	40%	[49]
Organosolv hardwood lignin	Nylon	40–60%	[50]
Softwood lignin from soda cooking process	PLA	20% and 40%	[52]
Alkali lignin and organosolv lignin	PLA	0.5–20%	[53]
Organosolv lignin	Polymer resins	5–15%	[54]
Softwood lignin from *P. radiata*	Polyhydroxybutyrate	20%	[55]

In an experiment where organosolv hardwood lignin was blended with nylon, the lignin was found to reinforce the thermoplastic matrix by increasing the stiffness of the structure, leading to increased 3D printability [50]. The melt viscosity was also reduced as a result of blending the lignin, which further increases 3D printability. The proposed mechanism for the bonding in the composite was lignin domains forming hydrogen bonds with the thermoplastic matrix. In addition, a study shows that 20–40% lignin can be used with PLA to form a matrix material for 3D printing. In the study, characterization techniques including thermogravimetric analysis, X-ray diffraction, and scanning electron microscope indicated that lignin is a nucleating agent that increases the crystallization of PLA [52]. This composite material resulted in great extrudability and flowability; it was also observed that lignin did not agglomerate. In another study, alkali lignin and organosolv lignin were acetylated to improve the compatibility with PLA to form a composite [53]. Even though the lignin decreased the crystallization behavior of PLA, it drastically improved the thermal stability of PLA and increased the elongation at break. The acetylated lignin was observed to prevent PLA from undergoing hydrolytic degradation.

Besides combining lignin with PLA, other studies have shown the diverse types of plastics that lignin can be blended with to construct effective 3D printing materials. For instance, authors of a study constructed photoactive acrylate resins used for 3D printing from up to 15 wt % acylated organosolv lignin, resin bases, a reactive diluent, and other compounds [54]. Figure 6 shows the reaction used for lignin acylation to construct resins. Even though the resulting lignin resin had decreased thermal stability compared to the commonly available resin, the resin exhibited characteristics that were favorable for 3D printing. The resin had increased ductility and resulted in 3D prints that were uniformly fused, high-resolution, and tough.

Additionally, a study reports that 20 wt % biorefinery lignin from *P. radiata* combined with polyhydroxybutyrate (PHB) improved the surface quality and reduced the shrinkage of the 3D printing material compared to the PHB alone [55]. Figure 7 below is a flow chart depicting how lignin can be extracted from softwood chips and combined with PHB to construct an improved 3D printing material. The presence of lignin was reported to reduce wrapping by values between 38% and 78% when compared to PHB, used as the sole printing material. This reduction of shrinkage was possibly due to the complex bulk lignin structure in addition to the lack of interfacial tension between the lignin and the polymer matrix.

Figure 6. Reaction to synthesize acylated lignin for the construction of resin. Reprinted with permission from [54]. Copyright (2018) American Chemical Society.

Figure 7. Flow chart showing 3D printed product constructed from high-temperature mechanical pretreatment (HTMP)-derived lignin and polyhydroxybutyrate (PHB). Reprinted from reference [55], with permission from Elsevier.

6. Perspectives

In the past decade, there have been more investigations conducted on using lignin to construct medical materials, electrochemical energy materials, and 3D printing composites. From the limited technoeconomic assessments conducted, it is difficult to predict the economic trajectory of lignin valorization. Since lignin valorization is an emerging concept, production costs may be higher than conventional processes [3]. According to a technoeconomic assessment analyzing catechol production from lignin, the lignin depolymerization and product separation stages accounted for a total of 55% [3]. For the production of lignin micro- and nanoparticles, their investment cost was 160 million USD, with an atomizer and separation system costing roughly 65% of the capital investment [56]. Assuming Kraft lignin and lignosulfonates are used, manufacturing costs can vary between 870 and USD 1170 per ton. The production capacity of around 70% of North American Kraft pulp mills is restricted by the operation of recovery boilers. Another challenge is finding the optimal membrane used for the Kraft process that will result in the highest lignin yield and purity [57]. Factors to consider for future scale-up efforts would be the size of the market, price volatility of the desired product [3], and the cost of feedstock in addition to solvents [56]. Lignin commercialization is likely with increased international collaboration, as this will mobilize resources and facilitate the development of innovative pathways [57].

Materials constituting lignin and its derivatives have potential in the medical sector due to their versatility as a wound dressing, pharmaceutical drug, drug delivery system, and drug removal material. However, it is difficult for lignin to be absorbed by the human's small intestine or stomach because of its high molecular weight and polydispersity [27]. There have also been studies suggesting that lignin can adsorb toxins, cholesterol, and surfactants, which can negatively impact human health [27]. Much of the research conducted is still at the proof of concept stage and pharmaceutical drugs will need to go through clinical trials before determining their efficacy. Regulatory bodies, like the US Food and Drug Administration, will need to approve the medical products even before they can be commercialized [58]. Clinical trials are time-consuming and costly; thus, there needs to be sufficient confidence in the research done and the viability of the production process before pursuing clinical trials.

While lignin is cheap and eco-friendly as a feedstock material for battery and 3D printing materials, a significant challenge is that it is hard to control or isolate lignin's complex recalcitrant structures. Even when fractionation is employed to isolate a lignin structure, fractionation often results in a mixture of products [59,60]. The electrochemical performance of bio-based battery materials depends on pore features, surface chemistry, and the structure of the lignin raw material [2]. Even though lignin is a relatively cheap starting material, it is still challenging to produce high performing carbon materials for supercapacitors and batteries. Since some lignins cannot stabilize easily, they convert to carbon nanofibers and consequently their morphologies are difficult to control; this makes high-performing lignin-based electrodes difficult to achieve [61]. Furthermore, lignin reaction mechanisms are not well understood [2].

Incorporating lignin in 3D printing materials decreases the reliance on typical petroleum-based plastic 3D printing materials. Doing so not only makes the 3D printing materials more environmentally friendly, but it also reduces costs as lignin is much cheaper than typical 3D printing materials [62]. Additionally, a myriad of materials can not only be created by 3D printing, but they can also be customized. This is of interest in the medical sector, where wound dressings can be 3D printed and customized to best suit the patients' needs. While lignin and other materials are being developed as 3D printing materials, they will most likely not replace traditional 3D printing materials including plastics [62]. Studies conducted so far in this current review suggest that lignin can be used to enhance the properties of the lignin–plastic composite 3D printing material but will not be the main feedstock material for 3D printing.

Even though the technoeconomic assessments of lignin valorization present several challenges, lignin's versatility has attracted significant efforts to valorize lignin. To illustrate lignin's versatility as a renewable material, Figure 8 shows the number of publications and citations searched in the

Web of Science with the keywords: "lignin application." As presented, the number of publications and citations have substantially increased over the past decade, indicating that significant efforts have been made in lignin valorization. Considering that there has been increased market demand for renewable materials in a wide range of industries, innovative lignin applications will continue to be developed indefinitely.

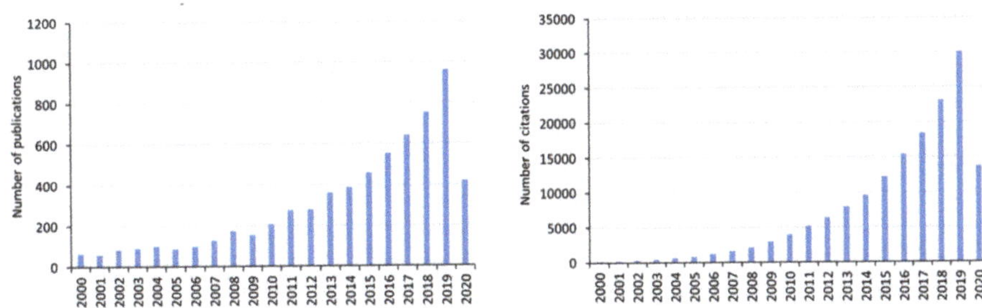

Figure 8. Number of publications (**left**) and citations (**right**) searched in the Web of Science with the keywords "lignin application" (as of May 30th 2020).

Author Contributions: Conceptualization, O.Y. and K.H.K.; Methodology, O.Y. and K.H.K.; Writing—Original draft preparation, O.Y.; Writing—Review and editing, O.Y. and K.H.K.; Supervision, K.H.K. All authors have read and agreed to the published version of the manuscript.

Funding: This research was funded by the Korea Institute of Science and Technology–The University of British Columbia Biorefinery onsite laboratory project, grant number 2E30740.

Conflicts of Interest: The authors declare no conflict of interest.

References

1. Cai, C.M.; Zhang, T.; Kumar, R.; Wyman, C.E. THF co-solvent enhances hydrocarbon fuel precursor yields from lignocellulosic biomass. *Green Chem.* **2013**, *15*, 3140–3145. [CrossRef]
2. Wu, X.; Jiang, J.; Wang, C.; Liu, J.; Pu, Y.; Ragauskas, A.; Li, S.; Yang, B. Lignin-derived electrochemical energy materials and systems. *Biofuels Bioprod. Biorefining* **2020**, *14*, 650–672. [CrossRef]
3. Bajwa, D.; Pourhashem, G.; Ullah, A.; Bajwa, S. A concise review of current lignin production, applications, products and their environmental impact. *Ind. Crop. Prod.* **2019**, *139*, 111526. [CrossRef]
4. Graglia, M.; Kanna, N.; Esposito, D. Lignin refinery: Towards the preparation of renewable aromatic building blocks. *ChemBioEng Rev.* **2015**, *2*, 377–392. [CrossRef]
5. Sahoo, S.; Seydibeyoğlu, M.; Mohanty, A.; Misra, M. Characterization of industrial lignins for their utilization in future value added applications. *Biomass Bioenergy* **2011**, *35*, 4230–4237. [CrossRef]
6. Seydibeyoğlu, M.Ö. A novel partially biobased PAN-lignin blend as a potential carbon fiber precursor. *BioMed Res. Int.* **2012**, *2012*, 598324.
7. Funaoka, M.; Matsubara, M.; Seki, N.; Fukatsu, S. Conversion of native lignin to a highly phenolic functional polymer and its separation from lignocellulosics. *Biotechnol. Bioeng.* **1995**, *46*, 545–552. [CrossRef] [PubMed]
8. Azadi, P.; Inderwildi, O.R.; Farnood, R.; King, D.A. Liquid fuels, hydrogen and chemicals from lignin: A critical review. *Renew. Sustain. Energy Rev.* **2013**, *21*, 506–523. [CrossRef]
9. Al Arni, S. Extraction and isolation methods for lignin separation from sugarcane bagasse: A review. *Ind. Crop. Prod.* **2018**, *115*, 330–339. [CrossRef]
10. Ragauskas, A.J.; Beckham, G.T.; Biddy, M.J.; Chandra, R.; Chen, F.; Davis, M.F.; Davison, B.H.; Dixon, R.A.; Gilna, P.; Keller, M. Lignin valorization: Improving lignin processing in the biorefinery. *Science* **2014**, *344*, 1246843. [CrossRef]
11. Chio, C.; Sain, M.; Qin, W. Lignin utilization: A review of lignin depolymerization from various aspects. *Renew. Sustain. Energy Rev.* **2019**, *107*, 232–249. [CrossRef]

12. Chen, H. *Lignocellulose Biorefinery Engineering: Principles and Applications*; Woodhead Publishing: Cambridge, UK, 2015.

13. Sjöström, E.; Westermark, U. Chemical composition of wood and pulps: Basic constituents and their distribution. In *Analytical Methods in Wood Chemistry, Pulping, and Papermaking*; Springer: Heidelberg, Germany, 1999; pp. 1–19.

14. Aro, T.; Fatehi, P. Production and application of lignosulfonates and sulfonated lignin. *ChemSusChem* **2017**, *10*, 1861–1877. [CrossRef] [PubMed]

15. Zhao, X.; Cheng, K.; Liu, D. Organosolv pretreatment of lignocellulosic biomass for enzymatic hydrolysis. *Appl. Microbiol. Biotechnol.* **2009**, *82*, 815. [CrossRef]

16. Zhang, K.; Pei, Z.; Wang, D. Organic solvent pretreatment of lignocellulosic biomass for biofuels and biochemicals: A review. *Bioresour. Technol* **2016**, *199*, 21–33. [CrossRef] [PubMed]

17. Kim, K.H.; Dutta, T.; Ralph, J.; Mansfield, S.D.; Simmons, B.A.; Singh, S. Impact of lignin polymer backbone esters on ionic liquid pretreatment of poplar. *Biotechnol. Biofuels* **2017**, *10*, 101. [CrossRef] [PubMed]

18. Kim, K.H.; Eudes, A.; Jeong, K.; Yoo, C.G.; Kim, C.S.; Ragauskas, A. Integration of renewable deep eutectic solvents with engineered biomass to achieve a closed-loop biorefinery. *Proc. Natl. Acad. Sci. USA* **2019**, *116*, 13816–13824. [CrossRef] [PubMed]

19. Ahmed, E.M. Hydrogel: Preparation, characterization, and applications: A review. *J. Adv. Res.* **2015**, *6*, 105–121. [CrossRef]

20. Laftah, W.A.; Hashim, S.; Ibrahim, A.N. Polymer hydrogels: A review. *Polym. Plast. Technol. Eng.* **2011**, *50*, 1475–1486. [CrossRef]

21. Yadollahi, M.; Namazi, H.; Aghazadeh, M. Antibacterial carboxylmethyl cellulose/Ag nanocomposite hydrogels cross-linked with layered double hydroxides. *Int. J. Biol. Macromol.* **2015**, *79*, 269–277. [CrossRef]

22. Asina, F.; Brzonova, I.; Kozliak, E.; Kubatova, A.; Ji, Y. Microbial treatment of industrial lignin: Successes, problems and challenges. *Renew. Sustain. Energy Rev.* **2017**, *77*, 1179–1205. [CrossRef]

23. Zhanga, Y.; Jiangb, M.; Zhanga, Y.; Caoa, Q.; Wanga, X.; Hana, Y.; Suna, G.; Lia, Y.; Zhoua, J. Novel lignin–chitosan–PVA composite hydrogel for wound dressing. *Mater. Sci. Eng. C* **2019**, *104*, 110002. [CrossRef] [PubMed]

24. Spasojević, D.; Zmejkoski, D.; Glamočlija, J.; Nikolić, M.; Soković, M.; Milošević, V.; Jarić, I.; Stojanović, M.; Marinković, E.; Barisani-Asenbauer, T.; et al. Lignin model compound in alginate hydrogel: A strong antimicrobial agent with high potential in wound treatment. *Int. J. Antimicrob. Agents* **2016**, *48*, 732–735. [CrossRef] [PubMed]

25. Lia, M.; Jianga, X.; Wangb, D.; Xua, Z.; Yanga, M. In situ reduction of silver nanoparticles in the lignin based hydrogel for enhanced antibacterial application. *Colloids Surf. B Biointerfaces* **2019**, *177*, 370–376. [CrossRef] [PubMed]

26. Domínguez-Robles, J.; Martin, N.K.; Fong, M.L.; Stewart, S.A.; Irwin, N.J.; Rial-Hermida, M.I.; Donnelly, R.F.; Larrañeta, E. Antioxidant PLA composites containing lignin for 3d printing applications: A potential material for healthcare applications. *Pharmaceutics* **2019**, *11*, 165. [CrossRef] [PubMed]

27. Gil-Chávez, J.; Gurikov, P.; Hu, X.; Meyer, R.; Reynolds, W.; Smirnova, I. Application of novel and technical lignins in food and pharmaceutical industries: Structure-function relationship and current challenges. *Biomass Convers. Biorefinery* **2019**, 1–17. [CrossRef]

28. Roopan, S.M. An overview of natural renewable bio-polymer lignin towards nano and biotechnological applications. *Int. J. Biol. Macromol.* **2017**, *103*, 508–514. [CrossRef]

29. Hasegawa, Y.; Kadota, Y.; Hasegawa, C.; Kawaminami, S. Lignosulfonic acid-induced inhibition of intestinal glucose absorption. *J. Nutr. Sci. Vitaminol.* **2015**, *61*, 449–454. [CrossRef]

30. Gordts, S.C.; Ferir, G.; D'huys, T.; Petrova, M.I.; Lebeer, S.; Snoeck, R.; Andrei, G.; Schols, D. The low-cost compound lignosulfonic acid (LA) exhibits broad-spectrum anti-HIV and anti-HSV activity and has potential for microbicidal applications. *PLoS ONE* **2015**, *10*, e0131219. [CrossRef]

31. Norikura, T.; Mukai, Y.; Fujita, S.; Mikame, K.; Funaoka, M.; Sato, S. Lignophenols Decrease Oleate-Induced Apolipoprotein-B Secretion in HepG2 Cells. *Basic Clin. Pharmacol. Toxicol.* **2010**, *107*, 813–817. [CrossRef]

32. Elangovan, S.; Afanasenko, A.; Haupenthal, J.r.; Sun, Z.; Liu, Y.; Hirsch, A.K.; Barta, K. From wood to tetrahydro-2-benzazepines in three waste-free steps: Modular synthesis of biologically active lignin-derived scaffolds. *ACS Cent. Sci.* **2019**, *5*, 1707–1716. [CrossRef]

33. Wahba, S.M.; Darwish, A.S.; Shehata, I.H.; Elhalem, S.S.A. Sugarcane bagasse lignin, and silica gel and magneto-silica as drug vehicles for development of innocuous methotrexate drug against rheumatoid arthritis disease in albino rats. *Mater. Sci. Eng. C* **2015**, *48*, 599–610. [CrossRef] [PubMed]

34. Figueiredo, P.; Lintinen, K.; Kiriazis, A.; Hynninen, V.; Liu, Z.; Bauleth-Ramos, T.; Rahikkala, A.; Correia, A.; Kohout, T.; Sarmento, B. In vitro evaluation of biodegradable lignin-based nanoparticles for drug delivery and enhanced antiproliferation effect in cancer cells. *Biomaterials* **2017**, *121*, 97–108. [CrossRef]

35. Farhat, W.; Venditti, R.; Mignard, N.; Taha, M.; Becquart, F.; Ayoub, A. Polysaccharides and lignin based hydrogels with potential pharmaceutical use as a drug delivery system produced by a reactive extrusion process. *Int. J. Biol. Macromol.* **2017**, *104*, 564–575. [CrossRef] [PubMed]

36. Żółtowska-Aksamitowska, S.; Bartczak, P.; Zembrzuska, J.; Jesionowski, T. Removal of hazardous non-steroidal anti-inflammatory drugs from aqueous solutions by biosorbent based on chitin and lignin. *Sci. Total Environ.* **2018**, *612*, 1223–1233. [CrossRef] [PubMed]

37. Gómez-Avilés, A.; Peñas-Garzón, M.; Bedia, J.; Rodriguez, J.; Belver, C. C-modified TiO$_2$ using lignin as carbon precursor for the solar photocatalytic degradation of acetaminophen. *Chem. Eng. J.* **2019**, *358*, 1574–1582. [CrossRef]

38. Camiré, A.; Espinasse, J.; Chabot, B.; Lajeunesse, A. Development of electrospun lignin nanofibers for the adsorption of pharmaceutical contaminants in wastewater. *Environ. Sci. Pollut. Res.* **2020**, *27*, 3560–3573. [CrossRef]

39. Zhu, J.; Yan, C.; Zhang, X.; Yang, C.; Jiang, M.; Zhang, X. A sustainable platform of lignin: From bioresources to materials and their applications in rechargeable batteries and supercapacitors. *Prog. Energy Combust. Sci.* **2020**, *76*, 100788. [CrossRef]

40. Temerov, F.; Belyaev, A.; Ankudze, B.; Pakkanen, T.T. Preparation and photoluminescence properties of graphene quantum dots by decomposition of graphene-encapsulated metal nanoparticles derived from Kraft lignin and transition metal salts. *J. Lumin.* **2019**, *206*, 403–411. [CrossRef]

41. Tang, P.-D.; Du, Q.-S.; Li, D.-P.; Dai, J.; Li, Y.-M.; Du, F.-L.; Long, S.-Y.; Xie, N.-Z.; Wang, Q.-Y.; Huang, R.-B. Fabrication and characterization of graphene microcrystal prepared from lignin refined from sugarcane bagasse. *Nanomaterials* **2018**, *8*, 565. [CrossRef]

42. Ding, Z.; Li, F.; Wen, J.; Wang, X.; Sun, R. Gram-scale synthesis of single-crystalline graphene quantum dots derived from lignin biomass. *Green Chem.* **2018**, *20*, 1383–1390. [CrossRef]

43. Lai, C.; Zhou, Z.; Zhang, L.; Wang, X.; Zhou, Q.; Zhao, Y.; Wang, Y.; Wu, X.-F.; Zhu, Z.; Fong, H. Free-standing and mechanically flexible mats consisting of electrospun carbon nanofibers made from a natural product of alkali lignin as binder-free electrodes for high-performance supercapacitors. *J. Power Sour.* **2014**, *247*, 134–141. [CrossRef]

44. Ago, M.; Borghei, M.; Haataja, J.S.; Rojas, O.J. Mesoporous carbon soft-templated from lignin nanofiber networks: Microphase separation boosts supercapacitance in conductive electrodes. *Rsc Adv.* **2016**, *6*, 85802–85810. [CrossRef]

45. Zhang, W.; Zhao, M.; Liu, R.; Wang, X.; Lin, H. Hierarchical porous carbon derived from lignin for high performance supercapacitor. *Colloids Surf. A Physicochem. Eng. Asp.* **2015**, *484*, 518–527. [CrossRef]

46. Gnedenkov, S.V.; Opra, D.P.; Sinebryukhov, S.L.; Tsvetnikov, A.K.; Ustinov, A.Y.; Sergienko, V.I. Hydrolysis lignin: Electrochemical properties of the organic cathode material for primary lithium battery. *J. Ind. Eng. Chem.* **2014**, *20*, 903–910. [CrossRef]

47. Chang, Z.-Z.; Yu, B.-J.; Wang, C.-Y. Influence of H2 reduction on lignin-based hard carbon performance in lithium ion batteries. *Electrochim. Acta* **2015**, *176*, 1352–1357. [CrossRef]

48. Culebras, M.; Geaney, H.; Beaucamp, A.; Upadhyaya, P.; Dalton, E.; Ryan, K.M.; Collins, M.N. Bio-derived Carbon Nanofibres from Lignin as High-Performance Li-Ion Anode Materials. *ChemSusChem* **2019**, *12*, 4516–4521. [CrossRef]

49. Nguyen, N.A.; Bowland, C.C.; Naskar, A.K. A general method to improve 3D-printability and inter-layer adhesion in lignin-based composites. *Appl. Mater. Today* **2018**, *12*, 138–152. [CrossRef]

50. Nguyen, N.A.; Barnes, S.H.; Bowland, C.C.; Meek, K.M.; Littrell, K.C.; Keum, J.K.; Naskar, A.K. A path for lignin valorization via additive manufacturing of high-performance sustainable composites with enhanced 3D printability. *Sci. Adv.* **2018**, *4*, eaat4967. [CrossRef]

51. Huang, J.; Liu, W.; Qiu, X. High performance thermoplastic elastomers with biomass lignin as plastic phase. *ACS Sustain. Chem. Eng.* **2019**, *7*, 6550–6560. [CrossRef]

52. Tanase-Opedal, M.; Espinosa, E.; Rodríguez, A.; Chinga-Carrasco, G. Lignin: A biopolymer from forestry biomass for biocomposites and 3D printing. *Materials* **2019**, *12*, 3006. [CrossRef]

53. Gordobil, O.; Egüés, I.; Llano-Ponte, R.; Labidi, J. Physicochemical properties of PLA lignin blends. *Polym. Degrad. Stab.* **2014**, *108*, 330–338. [CrossRef]

54. Sutton, J.T.; Rajan, K.; Harper, D.P.; Chmely, S.C. Lignin-containing photoactive resins for 3D printing by stereolithography. *ACS Appl. Mater. Interfaces* **2018**, *10*, 36456–36463. [CrossRef]

55. Vaidya, A.A.; Collet, C.; Gaugler, M.; Lloyd-Jones, G. Integrating softwood biorefinery lignin into polyhydroxybutyrate composites and application in 3D printing. *Mater. Today Commun.* **2019**, *19*, 286–296. [CrossRef]

56. Abbati de Assis, C.; Greca, L.G.; Ago, M.; Balakshin, M.Y.; Jameel, H.; Gonzalez, R.; Rojas, O.J. Techno-economic assessment, scalability, and applications of aerosol lignin micro-and nanoparticles. *ACS Sustain. Chem. Eng.* **2018**, *6*, 11853–11868. [CrossRef] [PubMed]

57. Fang, Z.; Smith Jr, R.L. *Production of Biofuels and Chemicals from Lignin*; Springer: Singapore, 2016.

58. Domínguez-Robles, J.; Larrañeta, E.; Fong, M.L.; Martin, N.K.; Irwin, N.J.; Mutjé, P.; Tarrés, Q.; Delgado-Aguilar, M. Lignin/poly (butylene succinate) composites with antioxidant and antibacterial properties for potential biomedical applications. *Int. J. Biol. Macromol.* **2020**, *145*, 92–99. [CrossRef] [PubMed]

59. Vinardell, M.P.; Mitjans, M. Lignins and their derivatives with beneficial effects on human health. *Int. J. Mol. Sci.* **2017**, *18*, 1219. [CrossRef]

60. Witzler, M.; Alzagameem, A.; Bergs, M.; Khaldi-Hansen, B.E.; Klein, S.E.; Hielscher, D.; Kamm, B.; Kreyenschmidt, J.; Tobiasch, E.; Schulze, M. Lignin-derived biomaterials for drug release and tissue engineering.figure. *Molecules* **2018**, *23*, 1885. [CrossRef]

61. Collins, M.N.; Nechifor, M.; Tanasă, F.; Zănoagă, M.; McLoughlin, A.; Stróżyk, M.A.; Culebras, M.; Teacă, C.-A. Valorization of lignin in polymer and composite systems for advanced engineering applications–a review. *Int. J. Biol. Macromol.* **2019**, *131*, 828–849. [CrossRef]

62. MarketsandMarkets. 3D Printing Materials Market—Global Forecast to 2024. Available online: https://www.marketsandmarkets.com/Market-Reports/3d-printing-materials-market-1295.html (accessed on 4 June 2020).

Article

Techno-Economic Evaluation of Hand Sanitiser Production Using Oil Palm Empty Fruit Bunch-Based Bioethanol by Simultaneous Saccharification and Fermentation (SSF) Process

Andre Fahriz Perdana Harahap [1], Jabosar Ronggur Hamonangan Panjaitan [2], Catia Angli Curie [1], Muhammad Yusuf Arya Ramadhan [1], Penjit Srinophakun [3] and Misri Gozan [1,*]

[1] Chemical Engineering Department, Faculty of Engineering, Universitas Indonesia, Depok 16424, Indonesia; andrefahriz25@gmail.com (A.F.P.H.); catiacurie@gmail.com (C.A.C.); yra.ramadhan@gmail.com (M.Y.A.R.)

[2] Chemical Engineering Program, Institut Teknologi Sumatera, Lampung 35365, Indonesia; jabosarronggur@gmail.com

[3] Chemical Engineering Department, Faculty of Engineering, Kasetsart University, Bangkok 10900, Thailand; fengpjs@ku.ac.th

* Correspondence: mgozan@che.ui.ac.id; Tel.: +62-857-811-76292

Received: 22 July 2020; Accepted: 26 August 2020; Published: 29 August 2020

Featured Application: The results of this work can be applied as the basis of consideration during the feasibility study step of plant design supposed to produce biomass-based bioethanol for commercial-scale hand sanitiser production. The results can contribute in the effort of converting agricultural waste into more valuable alternative product such as hand sanitiser which is urgently needed in global pandemic situation.

Abstract: Oil palm empty fruit bunch (OPEFB) is a potential raw material abundantly available for bioethanol production. However, the second-generation bioethanol is still not yet economically feasible. The COVID-19 pandemic increases the demand for ethanol as the primary ingredient of hand sanitisers. This study evaluates the techno-economic feasibility of hand sanitiser production using OPEFB-based bioethanol. OPEFB was alkaline-pretreated, and simultaneous saccharification and fermentation (SSF) was then performed by adding *Saccharomyces cerevisiae* and cellulose enzyme. The cellulose content of the OPEFB increased from 39.30% to 63.97% after pretreatment. The kinetic parameters of the OPEFB SSF at 35 °C, which included a μ max, ks, and kd of 0.018 h^{-1}, 0.025 g/dm^3, and 0.213 h^{-1}, respectively, were used as input in SuperPro Designer® v9.0. The total capital investment (TCI) and annual operating costs (AOC) of the plant were \$645,000 and \$305,000, respectively, at the capacity of 2000 kg OPEFB per batch. The batch time of the modelled plant was 219 h, with a total annual production of 32,506.16 kg hand sanitiser. The minimum hand sanitiser selling price was found to be \$10/L, achieving a positive net present value (NPV) of \$108,000, showing that the plant is economically feasible.

Keywords: bioethanol; economic analysis; hand sanitiser; oil palm empty fruit bunch (OPEFB); simultaneous saccharification and fermentation; SuperPro Designer®

---

1. Introduction

The World Health Organization (WHO) declared the outbreak of novel coronavirus SARS-CoV-2 as a global pandemic because of its ease of spread, severity it may cause and the lack of global precaution back then [1]. SARS-CoV-2 can be easily transmitted among humans through droplets, and human-coronaviruses can remain active on surfaces for up to nine days [2]. People may also be infected without showing severe symptoms. Practising cough and sneeze etiquette, as well as hand hygiene, is thus encouraged as a preventive measure. Following the WHO pandemic declaration, many countries then put in place strong restrictions and even lockdown to urge their citizens to limit physical contact and slow down the spread of Covid-19 as much as possible. However, as of June 2020, countries are beginning to relax the lockdown and restrictions because of economic demand. Since a vaccine for COVID-19 is yet to be developed, individual precautions must be practised in a more disciplined manner to prevent the spread of the virus. These have made hand sanitiser one of the most crucial items to have on hand.

Hand sanitiser helps the achievement of hand hygiene, especially in clinical settings or in conditions where water is unavailable. A study performed by Rai et al. has shown that single-use hand sanitiser packets are quicker to use than single-use moist towelettes. Thus, hand sanitisers are often preferred [3]. However, most hand sanitisers are not effective for non-enveloped viruses. Thus, despite being practical, hand sanitiser cannot fully replace hand washing, especially when the hand is visibly dirty or has come into contact with harmful chemicals [4–6].

Hand sanitisers can be classified into two groups: alcohol-based and alcohol-free sanitisers. Alcohol-based hand sanitisers are more common because of the ease of preparation, low cost and efficacy. The WHO has released a guide for the local production of alcohol-based hand-rub formulations. Generally, the alcohol used can be ethanol, isopropyl alcohol, or n-propanol. An alcohol content of 60–95% *v/v* is required for alcohol-based sanitisers to kill microbes effectively. Alcohols render microbes ineffective by damaging their lipid membranes and/or denaturing the proteins. The WHO-recommended formulations contain either 80% *v/v* of ethanol or 75% *v/v* of isopropyl alcohol with 1.450% *v/v* glycerol, 0.125% *v/v* hydrogen peroxide and water. Compared to isopropanol, ethanol is less irritant to the skin and more effective in killing a broader range of microbes [5–7].

Ethanol is commonly produced either from petroleum feedstock or from a sugar/starch feedstock, which is referred to as first-generation bioethanol. Neither is sustainable because petroleum is non-renewable while using sugar or starch as a feedstock interferes with food supplies and the utilisation of fertile land. Hence, the use of lignocellulosic feedstock, second-generation bioethanol, offers greater potential in terms of sustainability. This is because lignocellulosic materials are derived from various organic waste products and residues, including agricultural, forestry and household waste. Lignocellulose generally contains cellulose (40–60%), hemicellulose (20–40%) and lignin (10–25%) with the exact content depends on the biomass source and harvest time. To produce ethanol, lignocellulosic feedstocks must undergo four basic steps: pretreatment, hydrolysis, fermentation and purification. The four steps may be assembled using the separate hydrolysis and fermentation (SHF) process, the simultaneous saccharification and fermentation (SSF) process, the simultaneous saccharification and co-fermentation (SSCF) process or consolidated bioprocessing (CBP). In SHF, each step stands alone. Thus, each can be performed under its optimum conditions. However, the capital cost of the SHF is high. In SSF, the sugar released from the saccharification process is directly fermented in the same reactor, limiting the risk of inhibition. SSF is currently the process used most often. Yet, saccharification and fermentation have different optimum conditions. Thus, it is still a challenge to perform both processes optimally in one reactor. SSCF differs from SSF in that it allows for the simultaneous fermentation of all the sugars produced from cellulose and hemicellulose. CBP, in turn, is different from the other processes because it uses only one type of microorganism to produce the enzymes for hydrolysis and to obtain fermentation. This process offers the most efficient process as it combines enzyme production, enzymatic hydrolysis and fermentation in one pot. Yet, the development of the CBP process is dependent upon the ability to engineer the appropriate microbes [8–10].

Many studies have been conducted to assess the production of bioethanol from various lignocellulose sources, such as from sugarcane bagasse using the SSF method and pretreated with white-rot fungi, from durian skin using the SSF method and cell encapsulation of *S. cerevisiae*, from paper waste using the SSF method, from corncobs using the SSCF method and scaled up to demo scale and from sugarcane bagasse using the CBP process [11–16]. Because of the complexity of the process, to date, lignocellulosic ethanol is still more expensive to produce compared to first-generation bioethanol. The productions of second-generation bioethanol are still covering less than 3% of total global bioethanol production where they are limited only at negligible amount in some demo plants around the world that work industrially but are not yet economically feasible [17,18]. One way of tackling this challenge is to produce various by-products in lignocellulosic ethanol plants. Rosales-Calderon and Arantes have reviewed several biochemicals with a minimum technology-readiness level of eight that can be produced alongside lignocellulosic ethanol [9]. The production of hand sanitisers can also be an option because of the ease of production and the increasing demand.

Oil palm empty fruit bunches (OPEFBs) are solid waste lignocellulosic biomass products of the oil palm industry, consisting of cellulose, hemicellulose and lignin. As much as 1.1 tonnes of OPEFBs can be produced from processing 1 tonne of palm oil. In 2011, the global production of OPEFBs amounted to 14.5 million tonnes, of which half were produced in Indonesia [19–21]. Their cellulose, hemicellulose and lignin contents give OPEFBs the potential to be converted into various biochemicals, such as bioethanol, furfural, formic acid and levulinic acid. Sahlan et al. have examined bioethanol production from OPEFBs using *Rhizopus oryzae* encapsulated using calcium alginate [14]. Panjaitan and Gozan have investigated formic acid production from the acid catalysed hydrolysis reaction of OPEFBs; Harahap et al. have further optimised the production process using the response surface methodology (RSM) method [22,23]. In addition, Gozan et al. have produced levulinic acid and furfural from OPEFBs and have evaluated its production kinetics [24].

As a demand and sustainability, palm oil serves many tasks such as excellent source of biomass, self-sufficient energy in processing, effective carbon sink, positive contribution to energy balance, sustainable practices and zero burning. The Indonesian Sustainable Palm Oil (ISPO) Certification Scheme is introduced by the Government of Indonesia as a mandatory requirement for all oil palm growers and mills to enhance the competitiveness of Palm Oil in the global market. For Malaysia, The Malaysian Sustainable Palm Oil (MSPO) Certification Scheme is the national scheme in Malaysia for oil palm plantations, independent and organised smallholdings, and palm oil processing facilities to be certified against the requirements of the MSPO Standards. The MSPO Certification Scheme provides the general principles for plantations and palm oil processing facilities to ensure that the palm oil products are produced in a responsible and sustainable manner [25].

Process simulation can be used to gain a better understanding of particular processes and their behaviours in real life. Simulation can be used to conduct experiments to evaluate behaviours or alternatives to chemical process systems [26]. To simulate a process, process engineering software needs specific scientific abilities, including, but not limited to, the ability to accurately describe the physical properties of pure components and complex mixtures, the ability to model a large variety of reactors and unit operations, and the computational and numerical techniques needed to solve a wide variety of different mathematical equations [27].

After thoroughly analysing a chemical process system, an economic analysis is an important step in developing better knowledge about that process. There are several parameters used to measure the investment performance of the production of chemical products, including, but not limited to, the internal rate of return (IRR), the net present value (NPV), and the payback period (PBP). The net present value is the difference between the present value of cash inflows and outflows used to analyse the profitability of the project. The internal rate of return is a value representing the profitability potential of the project as a form of discount rate that reduces the NPV of the project to zero. The payback period is the amount of time needed to recover the initial investment from the project. The rate of investment (ROI) is a percentage showing the efficiency of a project calculated from the net profit divided by the total cost of investment [28].

Advances in computational technology and improvements in modelling techniques have enabled easier access to quantitative predictions of complex systems' behaviours; in particular, mass and energy balances could, with further implementation, be used to accurately design unit operations. Process simulation has been commonly utilised in real-world chemical engineering industries to perform system analysis and synthesis [29]. Some of the leading commercial process simulation software packages today include Aspen Plus®, Aspen HYSYS®, CHEMCAD and PRO/II with PROVISION. The above simulators have been designed to model primarily continuous processes and their transient behaviour for process control purposes. Most bioprocess and biorefinery products, however, are produced in batch and semi-continuous modes. Such processes are best modelled with batch process simulators that account for time-dependencies and the sequencing of events. It is not difficult to switch between simulators once the principles of process simulation have been understood by the engineers [30]. SuperPro Designer® v.9.0. (Intelligen, Inc., Scotch Plains, NJ, USA, 2019) is one of the simulators specifically developed to simulate batch process systems. SuperPro Designer® also has the capability to perform comprehensive economic analyses, which is not limited to calculating total capital investment, NPV, IRR and PBP but can also perform detailed calculations of operational expenditure (OPEX) by enabling the customisation of staffing, utility pricing, and process scheduling. Therefore, SuperPro Designer® is perfect for the simulation needed in this study.

Until now, there have been no articles yet addressing the economic feasibility to produce value-added product like hand sanitiser, which is globally needed in the current pandemic situation, from under-utilised resource like OPEFBs. This study aims to provide an alternative application for second-generation bioethanol, combining both laboratory and modelling studies. A hand sanitiser plant utilising OPEFB as its feedstock was designed, and the kinetic parameters of the production of bioethanol from OPEFB using the SSF method were analysed. The obtained parameters were then used in the designed process simulation using SuperPro Designer® to evaluate the economic viability of this hand sanitiser plant.

2. Materials and Methods

2.1. Kinetic Evaluation of Fermentation

2.1.1. Preparation of Materials

The OPEFB used as the raw material for bioethanol production in this study was kindly received from the Indonesian government-owned palm oil plantation company PTPN (PT Perkebunan Nusantara) VIII Kertajaya, located in Banten, Indonesia. Before being used, the OPEFB was first reduced in size by cutting and grinding it in a laboratory-scale wood grinder (PT. Enerba Teknologi, Tangerang Selatan, Banten, Indonesia). The obtained OPEFB fibre was sifted through a 40-mesh stainless steel wire filter in order to obtain a uniform size of raw material, as shown in Figure 1. The OPEFB fibre was then washed under clean running water to remove the attached dirt and finally dried in an oven at 80 °C for 24 h. Cellulase enzymes from *Trichoderma reesei* ATCC 26921 and yeast *Saccharomyces cerevisiae* Type-1 were purchased from Sigma Aldrich (St. Louis, MO, USA). Both the cellulase and the yeast were stored in a refrigerator at 10 °C before being used. All other analytical-grade chemical reagents, like citrate buffer and sodium hydroxide, were purchased from Merck (Darmstadt, Hesse, Germany).

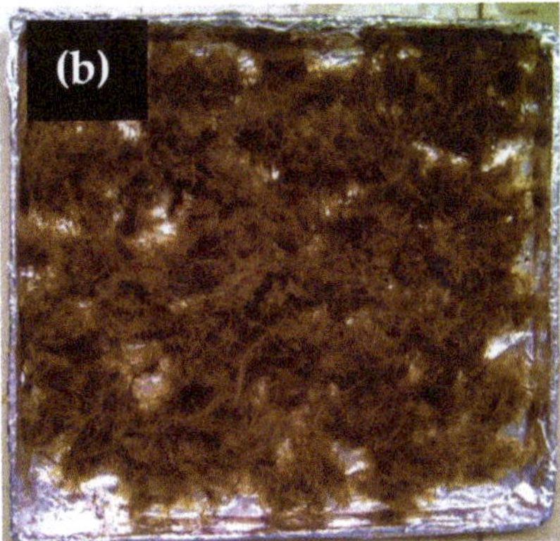

Figure 1. OPEFB before (**a**) and after (**b**) being ground as raw material for bioethanol production.

2.1.2. OPEFB Pretreatment and Analysis

OPEFB that will be used as a raw material for SSF must first be pretreated to remove its lignin content, thus enhancing the accessibility of the cellulase enzyme to the cellulose for further bioethanol conversion from glucose [31]. In this study, an alkaline pretreatment method using sodium hydroxide as the solvent was applied, according to Zulkiple et al. [32]. Pretreatment was performed by adding 30 g of the OPEFB sample into a 10% sodium hydroxide solution with a 1:10 solid-to-liquid ratio. The mixture was then heated to 120 °C for 2 h in an autoclave reactor (CV. Pugar Mandiri Teknik, Bandung, West Java, Indonesia) and cooled down at room temperature. The pretreated OPEFB was separated from the solvent, known as black liquor, using a stainless-steel wire filter. It was then washed with clean hot water to remove any remaining sodium hydroxide until neutrality was attained and dried in an oven at 95 °C for 24 h. The pretreated OPEFB was then analysed for its cellulose, hemicellulose, lignin and ash contents according to SNI 0444:2009, SNI 14-1304-1989, SNI 0492:2008 and SNI 0442:2009, respectively [33–36]. At this stage, the OPEFB was ready to be used for bioethanol production using the SSF method.

2.1.3. Simultaneous Saccharification and Fermentation

The SSF medium consisted of 15 g of pretreated OPEFB, 200 mL of 0.05 M citrate buffer (pH 4.8), 1 mL of cellulase enzyme and 5 g of dry yeast. The mixture of OPEFB and citrate buffer was first sterilised at 121 °C for 20 min in an autoclave. The mixture was then cooled down to room temperature and added to both the cellulase enzyme and the dry yeast. The SSF was performed in an Erlenmeyer flask (DWK Life Sciences GmbH, Mainz, Germany) closed tightly with aluminium foil. The flask was then incubated in an orbital shaker (Thermo Fisher Scientific, Waltham, MA, USA) at 150 rpm for 96 h. These experiments were conducted for temperature variations of 30 °C, 32 °C and 35 °C to give sufficient data for the determination of the kinetic parameters. Samples were taken at 24, 48, 72 and 96 h to analyse the concentrations of glucose, yeast cells and bioethanol in the fermentation broths.

2.1.4. Analysis of Fermentation Products

The fermented cell and glucose concentrations were analysed using optical density and dinitrosalicylic acid (DNS) methods. The optical density and DNS analyses were performed using UV-vis spectroscopy (UV-M90, BEL Engineering srl ©, Monza, MB, Italy). The DNS analysis used a 575 nm wavelength; the blank solution was a DNS buffer citrate solution. The optical density analysis used a 600 nm wavelength; the blank solution was a citrate buffer solution. The bioethanol analysis was conducted using a SUPELCOWAX-10 gas chromatography analyser (Supelco Inc., Bellefonte, PA, USA) at 50 °C.

2.1.5. Reaction Constants Determination

The kinetic model of bioethanol production used in this study was chosen in accordance with Fogler (2004), under the assumption that there is no mass-transfer effect and that the influence of bioethanol as an inhibitor can be ignored [37]. The kinetic equation can, therefore, be simplified to:

- The kinetic equation for yeast cell formation:

$$\frac{dC_C}{dt} = \left(\mu\,max\,\frac{C_C \cdot C_S}{ks + C_S} - kd \cdot C_C \right) \tag{1}$$

- The kinetic equation for product formation (bioethanol):

$$\frac{dC_C}{dt} = Y_{\frac{p}{C}} \cdot \mu\,max\,\frac{C_C \cdot C_S}{ks + C_S} \tag{2}$$

- The kinetic equation for the residual substrate:

$$\frac{dC_s}{dt} = Y_{\frac{s}{C}} \cdot \left(\mu\,max\,\frac{C_C \cdot C_S}{ks + C_S} \right) - m \cdot C_C \tag{3}$$

From the equations, it is necessary to determine the four reaction rate constants ($\mu\,max$, ks, kd and m). Experimental data were obtained for the cell, glucose and bioethanol concentrations over time, and the reaction rate constants could then be predicted by minimising the sum of the square error values between the experimental data and the prediction data that was formulated:

$$SS = \sum_i \sum_j \left([j]_{i\,exp} - [j]_{i\,predicted} \right)^2 \tag{4}$$

The sum of the square error value can be determined using fminsearch optimisation (the Nedler-Mead method) and MATLAB software (MATLAB 9.6, MathWorks, Natick, MA, USA, 2019).

2.2. Hand Sanitiser Plant Design

SuperPro Designer® v.9.0 was used to conducting the process simulation and economic assessment of this hand sanitiser plant. Process design should always be completed first in order to specify a step-by-step process as well as to determine the equipment that will be involved in the simulation. At this stage, users also input specific parameters and operating conditions directly in the software for each involved piece of equipment. To model this hand sanitiser plant, a capacity of 2000 kg per batch of OPEFB was determined. The plant is also assumed to be constructed inside a palm oil mill area to reduce the transportation cost associated with the OPEFB.

The first section of this hand sanitiser plant is the pretreatment section. In this section, OPEFB obtained from the palm oil mill is first reduced in size by a grinding machine. This equipment grinds the raw material from the palm oil mill into 40-mesh sized OPEFB for 1 h. The size-reduced OPEFB is then transferred into a vertical-on-legs tank, where the alkaline pretreatment takes place at 120 °C for 60 min. The OPEFB is mixed with a 10% sodium hydroxide solution with a 1:10 solid-to-liquid ratio by means of an adjustable mixer before entering this tank. The mixture is then cooled in a cooler for 60 min from 120 °C to 25 °C before being washed. During the washing process, the OPEFB is washed with 0.02 m³/kg of water for 30 min. This washing process is assumed to remove 60% of the water and 100% of the sodium hydroxide as aqueous waste known as black liquor. Some hemicellulose, lignin, and ash are also removed in this process, according to the pretreatment compositional analysis results.

The second section of the plant is devoted to medium preparation and SSF. The pretreated OPEFB from the previous washing process is mixed with a citrate buffer (0.380% citric acid, 0.720% sodium citrate and water) at 1:13.13 liquid-to-solid ratio by means of an adjustable mixer. The mixture is then sterilised using a heat steriliser at 120 °C for 2 h to avoid contamination from other organisms and cooled back down to 35 °C. Cellulase and yeast at 0.5% and 1% of the mass ratio of the entering feed, respectively, are gradually added to the mixture by means of two serial adjustable mixers. The mixture is then finally ready to be transferred into a batch vessel fermenter operating at 35 °C and equipped with a jacket for chilled water flow. SSF takes place in this fermenter, with three reactions occurring simultaneously. Equation (5) is related to the saccharification reaction. Equations (6) and (7) are related to ethanol production and cell formation, respectively. The batch time for SSF in this fermenter was set in such a way as to obtain the highest amount of bioethanol.

$$1.00 \text{ cellulose } + \ 13.13 \text{ water } \rightarrow 2.21 \text{glucose} \tag{5}$$

$$1.00 \text{ glucose } \rightarrow 2.09 \text{ carbon dioxide } + \ 1.91 \text{ethyl alcohol} \tag{6}$$

$$0.56 \text{ glucose } \rightarrow 0.45 \text{ carbon dioxide } + \ 1.11 \text{ water } + \ 0.33 \text{ yeast} \tag{7}$$

Reaction rate constants obtained from the optimisation results will be used as values to be input in the SuperPro Designer® v9.0 Software. The kinetics parameters data input can be seen in Table 1 below.

Table 1. Kinetics parameters data used as inputs in SuperPro Designer®.

Reaction	Kinetics Parameters Input
Glucose → Yeast cell	$\dfrac{dC_C}{dt} = \left[\mu\,\text{max}\left[1 - \dfrac{Cp}{Cp^{*}}\right]\dfrac{Cs}{ks+Cs} - kd\right]Cc$ Input: $\alpha = 1$ $\mu\,\text{max} = \mu\,\text{max}$ $(S1 - \text{Term}) = 1$ $(S2 - \text{Term}) = \dfrac{Cs}{ks+Cs}$ $\beta = -kd$ $(B - \text{Term}) = Cc$
Glucose → Ethanol	$\dfrac{dC_C}{dt} = \left[Y_{\frac{P}{C}} \cdot \mu\,\text{max}\left[1 - \dfrac{Cp}{Cp^{*}}\right]\dfrac{Cs}{ks+Cs}\right]Cc$ Input: $\alpha = Y_{\frac{P}{C}}$ (ethanol mass/cell mass) $\mu\,\text{max} = \mu\,\text{max}$ $(S1 - \text{Term}) = 1$ $(S2 - \text{Term}) = \dfrac{Cs}{ks+Cs}$ $\beta = 0$ $(B - \text{Term}) = Cc$
Glucose → Glucose(residue)	$\dfrac{dC_s}{dt} = \left[Y_{\frac{S}{C}} \cdot \left(\mu\,\text{max}\left[1 - \dfrac{Cp}{Cp^{*}}\right]\dfrac{Cs}{ks+Cs}\right) - m\right]Cc$ Input: $\alpha = Y_{\frac{S}{C}}$ $\mu\,\text{max} = \mu\,\text{max}$ $(S1 - \text{Term}) = 1$ $(S2 - \text{Term}) = \dfrac{Cs}{ks+Cs}$ $\beta = -m$ $(B - \text{Term}) = Cc$

The final section of the plant handles purification and hand sanitiser formulation. The purification is started with filtration for 2 h in a plate and frame filter, which is set to remove almost all solid fractions from the fermentation broth. The liquid fraction is then transferred into the first distillation column, called a beer column, for 6 h to separate the ethanol from the fermentation medium. Most of the water and fermentation medium are removed as bottom products. At the same time, concentrated ethanol is transferred to the second distillation column, called a rectifying column, to increase the ethanol concentration up to 96%. The product is then cooled to 25 °C and transferred into a blending tank, where 3% hydrogen peroxide, 98% glycerol and water are added to fulfil the hand sanitiser requirements provided by the WHO. The mixture is agitated for 1 h to obtain a homogenous hand sanitiser formulation and is then transferred out as a final product. The complete process flow diagram for the described hand-sanitiser production process as formulated by SuperPro Designer® v9.0 can be seen in Figure 2 below.

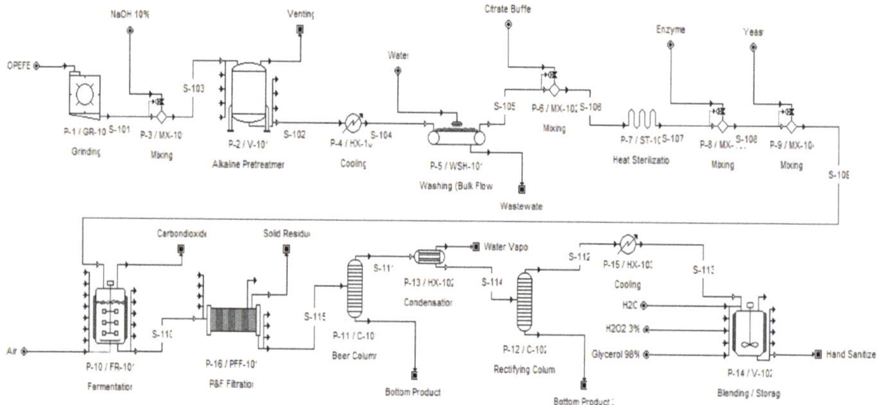

Figure 2. Complete process flow diagram of hand sanitiser production from OPEFB-based bioethanol in SuperPro Designer® v9.0.

The economic evaluation of this hand sanitiser plant was focused on the internal rate of return (IRR), the net present value (NPV), the payback period (PBP), the return on investment (ROI) and the gross margin of the operation. The calculations of those parameters were performed automatically by SuperPro Designer® v9.0. Before performing the economic evaluation, several terms should be assigned to this hand sanitiser plant, as can be seen in Table 2 below.

Table 2. Assigned terms for hand sanitiser plant's economic evaluation.

Parameters	Value
Year of analysis	2020
Construction period	30 months
Start-up period	4 months
Project lifetime	15 years
Interest rate	7.0%
OPEFB capacity	2000 kg/batch
Labour wage	$0.92/h

3. Results and Discussion

3.1. Kinetic Parameters of Fermentation

The compositional analysis showed that the un-pretreated OPEFB used in this study contained 39.0% cellulose, 29.8% hemicellulose, 22.8% lignin and 1.7% ash, according to Gozan et al. [24]. After being alkaline-pretreated, there were significant changes to the composition of the OPEFB, as can be seen in Table 3. The pretreated OPEFB contained 63.97% cellulose, 10.58% hemicellulose, 19.39% lignin and 1.51% ash. This result proved that the delignification of the OPEFB occurred during the pretreatment process. In general, alkaline pretreatment of biomass at elevated temperatures will dissolve hemicellulose and lignin effectively, as the degradation of hemicellulose occurs much faster than that of cellulose [38]. This phenomenon can be seen from the decreasing amounts of both hemicellulose and lignin, but not of cellulose, in the pretreated OPEFB. The high cellulose content in the pretreated OPEFB would be very beneficial for the hydrolysis reaction to produce glucose during SSF.

Table 3. Composition of OPEFB after NaOH pretreatment.

Parameters	Composition (%)
Ash	1.51
Lignin	19.39
Cellulose	63.97
Hemicellulose	10.58

The effect of the fermentation time on the cell concentration at various fermentation temperatures can be seen in Figure 3b. Based on Figure 3b, cell concentration tends to increase with longer SSF times. This shows that the process of cell formation is still at an exponential stage because of the availability of sufficient substrate during the reaction and of reaction conditions that support the process of cell growth. Cell growth phenomena such as the lag phase, stationary phase and death phase have not been seen in the SSF process; this could be due to the large range of times for sampling processes. The lag phase in the cell formation can be predicted at $t \leq 24$ h. The exponential phase can be seen according to Figure 3b from the 24th h to the 72nd h, and after the 72nd h, cell growth looks constant or enters the stationary phase. On the other hand, the death phase was not seen in this study. This could be due to the abundant amount of substrate, which is indicated by the large residue of OPEFBs at the end of the SSF reaction ($t = 96$ h). The death phase was also not seen in a study conducted by Amenaghawon et al. in which it was stated that the death phase is not seen in the cell growth phase because the substrate is still available for microorganisms to carry out metabolic processes [39].

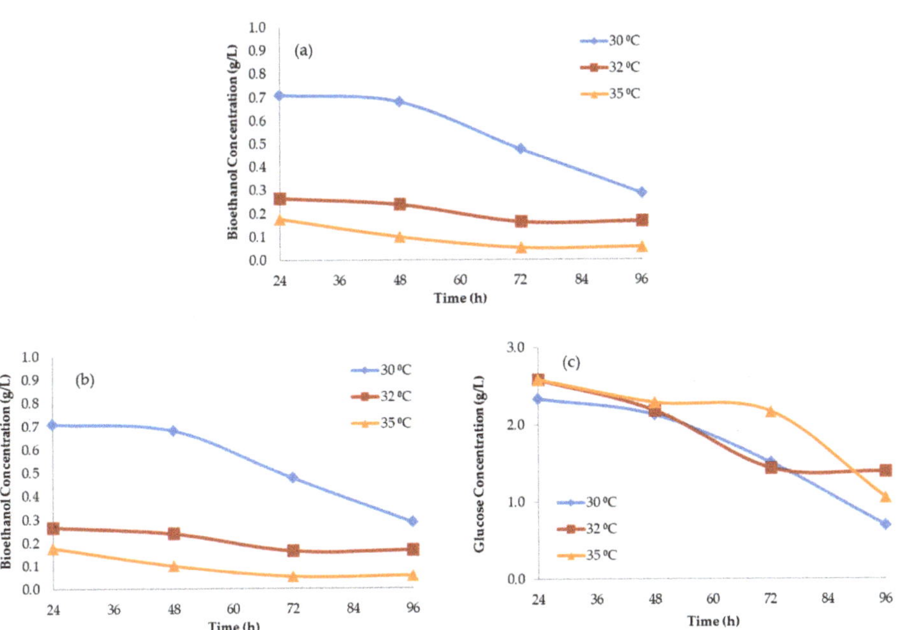

Figure 3. The concentration of bioethanol (**a**), yeast cell (**b**) and glucose (**c**) during SSF.

The effect of temperature on cell growth shows that the higher temperature of SSF produced higher microbial concentrations. This may have been due to the substrate that was used in this process.

The substrate used in this research was OPEFB, which must be broken down into glucose using the cellulase enzyme. The enzymatic hydrolysis process was generally carried out at a temperature of 50 °C ($T \geq 30$ °C). At 32 °C and 35 °C, the SSF process can cause the enzymatic hydrolysis of OPEFBs to be more effective and to produce more glucose than is produced at lower temperatures (30 °C). Glucose is used by yeast for cell growth.

The effect of the fermentation time on the glucose concentration at various fermentation temperatures can be seen in Figure 3c. Based on Figure 3c, glucose concentration tends to decrease with longer SSF times. This may occur because the glucose produced in enzymatic hydrolysis will be used directly by cells for growth processes and to produce bioethanol [39,40].

The decrease in glucose concentration showed the same tendency at each SSF reaction temperature. This indicates that in the SSF process, glucose that has been formed as a result of enzymatic hydrolysis will be used directly by yeast as a carbon source to form bioethanol. 30 °C was the temperature that showed the lowest glucose concentration at the end of the SSF reaction. This happened because, at this temperature, yeast cells make optimum use of glucose, which can be characterised by high ethanol concentrations.

The effect of the fermentation time on the bioethanol concentration at various fermentation temperatures can be seen in Figure 3a. Based on Figure 3a, the highest bioethanol concentration was 0.70 g/L. The longer reaction time of SSF produces lower bioethanol concentrations. The lower bioethanol concentration produced in this study was different from that produced in previous studies because, in general, the concentration of bioethanol will increase with longer fermentation times. The lower bioethanol concentration in this study could be due to the vaporisation of the ethanol.

The effect of temperature on bioethanol concentrations is quite significant. Figure 3a shows that a temperature of 30 °C with a 24-h reaction time produced bioethanol at around 0.70 g/L. At 32 °C and 35 °C with 24-h reaction time, the concentration of bioethanol dropped significantly to 0.26 g/L and 0.17 g/L, respectively. This might have occurred because higher reaction temperatures can make fermentation processes ineffective. High temperatures can affect cell transport activity and the occurrence of ribosomal and enzyme denaturation and can cause fluidity problems in cell membranes [41]. Yeast optimally ferments glucose into bioethanol at 30 °C. Based on data obtained from the laboratory, the bioethanol concentration was very small. Therefore, the influence of bioethanol as an inhibitor in the SSF kinetics equation can be ignored.

Based on Figure 4 above, the kinetic model optimisation of cell, glucose and bioethanol concentrations shows some results that are not optimal. This could be due to the composition of the raw materials used in this study. The raw material used in this study was OPEFB, which consists of cellulose, hemicellulose and lignin. The hemicellulose and lignin contents in the raw material could have inhibited the enzymatic hydrolysis process and the fermentation process. However, overall, the kinetics model that was used in this study is suitable for experimental data and can describe the SSF process.

From the results of the kinetic model optimisation with cell, glucose and bioethanol concentrations, various kinetic constants on the bioethanol reaction can be obtained. The kinetics constants for the bioethanol fermentation reaction can be seen in Table 4 below.

Table 4. Determined kinetic parameters for bioethanol production.

Kinetic Parameters	Temperature (°C)		
	30	32	35
μmax (h^{-1})	0.009	0.013	0.018
ks (g/dm^3)	0.004	0.010	0.025
kd (h^{-1})	0.009	0.009	0.213

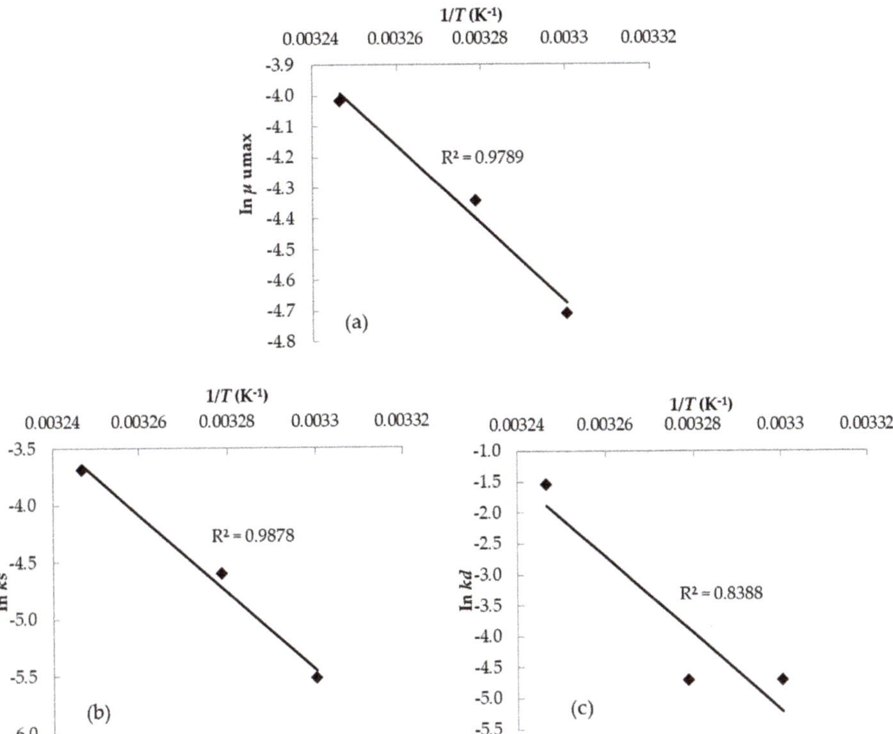

Figure 4. Arrhenius plot for SSF to determine μ max (**a**), *ks* (**b**) and *kd* (**c**).

Based on Table 4, it can be seen that higher temperatures will increase the value of various reaction constants, such as μ max (maximum specific growth rate), *ks* (Monod constant), *kd* (cell death rate constant) and m (maintenance cell constant). One parameter that influences cell growth is μ max. The effect of ethanol on cell growth can be seen through the μ max parameter. When the ethanol concentration rises, μ max tends to decrease. The same phenomenon is also seen in the previous research conducted by Amenaghawon et al. [39]. This indicates that bioethanol is an inhibitor of the cell growth process.

3.2. Process Simulation

The simulation result obtained using SuperPro Designer® v9.0 showed that fermentation time plays a significant role in determining the flowrate of products coming out of the fermenter when the kinetic parameters in Table 4 are applied. Figure 5 shows that the fermenter's output flowrate of bioethanol, glucose and yeast varied according to the fermentation time. The glucose flowrate decreased with increasing fermentation time as it was used as the substrate for bioethanol production. On the other hand, the bioethanol and yeast flowrates increased with increasing fermentation time, as they were both produced during fermentation. The glucose was completely consumed when the fermentation time was set to 192 h. At this time, the flowrate of produced bioethanol was at a maximum. Thus, the fermentation time for this hand sanitiser plant was set to 192 h to obtain the maximum bioethanol yield with the minimum batch time.

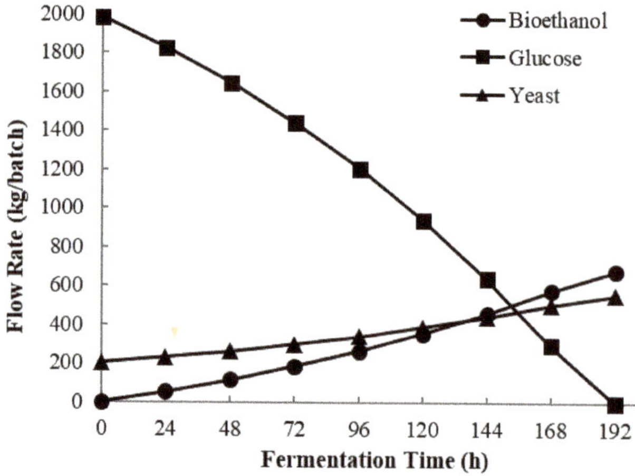

Figure 5. The relationship between fermentation time and flow rate of bioethanol, glucose and yeast leaving fermenter simulated by SuperPro Designer©. v9.0.

When the fermentation time was set to 192 h, the flowrate of produced bioethanol was 681 kg per batch, with a concentration of 3.4%. The plate and frame filtration could separate solid and liquid fractions of fermentation broth, resulting in an increased bioethanol concentration of 3.5% in the liquid output stream. The first distillation process in the beer column increased the bioethanol concentration to 79.3% with the remaining water. The second distillation process in the rectifying column finally increased the bioethanol concentration to 96.2%, as required by the WHO for use as a hand sanitiser raw material. At this point, the flowrate of bioethanol decreased to 651 kg due to some losses in the previous separation processes. Bioethanol was then mixed with the other raw materials, resulting in a final hand sanitiser product with a composition of 80.2% ethanol, 18.3% water, 1.4% glycerol and 0.1% hydrogen peroxide, as required by WHO standards. The overall results showed that this hand sanitiser plant design could produce 812.7 kg of hand sanitiser from 2000 kg of OPEFB for every batch.

The recipe scheduling information feature in SuperPro Designer® v9.0 for this hand sanitiser plant showed that the batch time and minimum cycle time (excluding equipment shared across batches and auxiliary equipment) were 219 and 194 h, respectively, with a total of 40 batches per year. This plant could, therefore, produce 32,506.16 kg of the total flow of hand sanitiser as the main product per year.

3.3. Cost and Economic Parameters

Investment performance measurement was conducted using several economic analyses, including the calculation of NPV, IRR, PBP and ROI. The calculations were performed using SuperPro Designer® v9.0 with several assumptions employed, based on the particulars of the Indonesian industry context, such as the number of holidays in a year, the labour price, the price of raw materials and the risk-free rate.

This section presents the estimation of the total capital investment cost (TCI), the annual operating cost (AOC) and the unit production cost of the hand-sanitiser product. The project is assumed to have a lifetime of 10 years and an annual operating time of 330 days (11 months). The depreciation of the capital investment was calculated using the straight-line method for a period of 10 years and a salvage value equal to 5% of the initial cost. The TCI of the plant was estimated according to the equipment purchase cost (EPC) following a well-established engineering methodology described in detail in the literature [42]. The equipment sizes were calculated using the simulation tool, based on the mass and energy balances. The EPC was estimated according to information from vendors, literature sources, and the SuperPro Designer® v9.0 equipment cost database. The total investment was estimated at $645,000; the breakdown of the TCI is presented in Table 5 below.

Table 5. Fixed capital estimate summary (2020 prices in $).

Total Plant Direct Cost (TPDC)	
1. Equipment purchase cost	101,000
2. Installation	33,000
3. Process piping	35,000
4. Instrumentation	41,000
5. Insulation	3000
6. Electrical	10,000
7. Buildings	46,000
8. Yard improvement	15,000
9. Auxiliary facilities	41,000
TPDC	325,000
Total Plant Indirect Cost (TPIC)	
10. Engineering	81,000
11. Construction	114,000
TPIC	195,000
Total Plant Cost (TPC = TPDC + TPIC)	
TPC	520,000
Contractor's Fee and Contingency (CFC)	
12. Contractor's Fee	26,000
13. Contingency	52,000
CFC	78,000
Direct Fixed Capital Cost (DFC = TPC + CFC)	
DFC	598,000

The calculation of the annual operating cost (AOC) was conducted to determine the influence of operating components on the product cost. The calculated AOC for this plant was $305,000; its breakdown is summarised in Figure 6. As shown in Figure 6, the largest percentage of the AOC is facility-dependent costs, which consist of capital depreciation, maintenance, insurance and overheads [43]. This can be explained, as the high value of the TCI and the relatively short lifetime of the plant lead to higher annual depreciation. The second highest cost contributor is utilities, which consist of standard electrical power, steam (low pressure and high pressure), and a large amount of cooling and chilled water, with a total annual utility cost of $95,749. The highest contributor to the utility cost is electricity, at $58,807 annually, or more than 61% of the utility cost. This is a result of the long period of fermentation that leads to high electricity consumption for processes such as agitation and temperature control. Raw materials, despite the low price of OPEFB, contribute to 22% of the AOC. The breakdown of raw material and utility costs can be seen in Table 6.

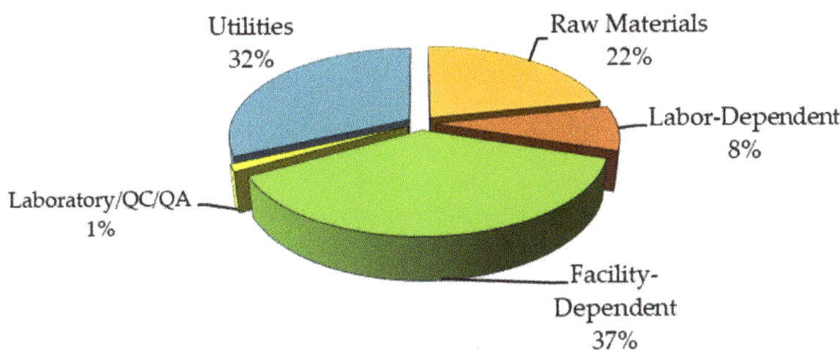

Figure 6. Annual operating cost breakdown.

Table 6. A detailed list of materials and utilities cost.

Material/Utility	Unit Cost ($)	Annual Amount	Reference Unit	Annual Cost ($)
Air	0.00	5,914,956	kg	0
Cellulase	2.50	4058	kg	10,146
Citric Acid	0.63	2866	kg	1791
Glycerol 98%	0.60	471	kg	283
H_2O_2 3%	0.40	1356	kg	542
OPEFB	0.01	80,000	kg	488
Sodium Citrate	0.76	5431	kg	4127
Sodium Hydroxide	0.43	80,000	kg	34,000
Water	0.01	4600	MT	46
Yeast	1.95	8158	kg	15,907
Standard Power	0.10	588,067	kW-h	58,807
Low Pressure Steam	3.34	1983	MT	6624
High Pressure Steam	5.78	1119	MT	6468
Cooling Water	0.05	186,277	MT	9314
Chilled Water	0.19	78,573	MT	14,536

One of the reasons why this project could be economically beneficial is the low cost of labour in Indonesia. The actual minimum wage of labour in Indonesia varies depending on location. In this study, we used the highest value of minimum wage in Indonesia for labour and blue-collar jobs, which is about $0.9/labour-h. The annual cost for operators as labour in this plant was found to be $24,996 with a total annual amount of 27,170 h, and it contributes to about 8% of the AOC.

After calculating the TCI and AOC, the unit production cost of the hand sanitiser product was then calculated at $9.37 per kg of the main product. Since hand-sanitiser products are sold by volume, we determined the product selling price in USD/L. We varied the product selling price to further analyse the investment performance and to obtain the best-selling price. The calculation was performed using SuperPro Designer®, and the parameters we focused on were IRR, PBP, gross margin, ROI and NPV. We wanted to obtain the selling price most feasible for penetrating the market in Indonesia, most likely to be economically beneficial and most able to fulfil the standard economic parameters of the market. The detailed data pertaining to the effect of product price variation can be seen in Table 7 below.

Table 7. Effect of hand sanitiser price on some economic parameters in this study.

Product Price (USD/L)	IRR (%)	PBP (Year)	Gross Margin (%)	ROI (%)	NPV (USD)
2	N/A	N/A	−295	−26	−1,824,000
4	N/A	N/A	−97	−14	−1,278,000
6	N/A	N/A	−32	−3	−732,000
8	0.4	10.9	1	9	−224,000
10	9.8	6.1	21	16	108,000
12	17.0	4.3	34	24	440,000
14	23.2	3.3	44	31	772,000
16	28.7	2.6	51	38	1,104,000

3.4. N/A: Not Available

Based on the data mentioned above, it can be concluded that a product selling price of $10/L is the minimum selling price able to fulfil all the given parameters. The hand sanitiser product with a $10/L price produces a 9.8% IRR, will break even at 6.1 years, and has an NPV of $108,000 at the end of the plant's lifetime. With a gross margin of 21%, the product has a lower value according to the average gross margin obtained by speciality chemical industries, which have gross margin values at 31.16% [44]. Therefore, we need to choose the product selling price that can fulfil this parameter. The price point of $12/L produces a gross margin value of 34% with a higher IRR and NPV compared to the $10/L price-point, with values of 17% and $440,000, respectively. It also has a relatively short payback period, at about 4.3 years. These economic parameters are from the market's point of view, with the average product selling price of hand sanitiser in Indonesia being $0.87 per 50 mL, or about $17.40/L, the product is economically competitive compared to the existing products in Indonesia.

4. Conclusions

This study explored the techno-economic evaluation of a hand sanitiser production plant using OPEFB-based bioethanol and the simultaneous saccharification and fermentation method. The kinetic parameters of bioethanol production using *Saccharomyces cerevisiae* in a laboratory-scale experiment were applied in SuperPro Designer® v9.0 to process simulation and economic assessment. The batch time of this hand sanitiser plant was 219 h, with a total of 40 batches per year. The total capital investment was calculated at $645,000 with a production capacity of 2000 kg per batch and a total of 32,506.16 kg of hand sanitiser as the main product per year. The total annual operating cost of this plant was found to be $305,000. The economic assessment performed by SuperPro Designer® v9.0 showed that the minimum selling price of the hand sanitiser was $10/L, which would result in a 9.8% IRR, a 21% gross margin, a 16% ROI, a $108,000 NPV and a PBP of 6.1 years. These results have shown that hand sanitiser production from OPEFB-based bioethanol is economically feasible and can be implemented at a tolerable price as an alternative application for second-generation bioethanol from lignocellulosic biomass in order to produce hygiene-related products.

Author Contributions: Conceptualization, M.G.; data curation, J.R.H.P.; formal analysis, A.F.P.H. and J.R.H.P.; investigation, J.R.H.P.; methodology, C.A.C.; project administration, M.Y.A.R.; resources, M.Y.A.R.; software, A.F.P.H.; supervision, P.S.; validation, P.S. and M.G.; visualization, A.F.P.H., C.A.C. and M.G.; writing—original draft, A.F.P.H.; writing—review and editing, P.S. and M.G. All authors have read and agreed to the published version of the manuscript.

Funding: We gratefully acknowledge the publication grant from Universitas Indonesia through Publikasi Terindeks International (PUTI) program Nr. NKB-1415/UN2.RST/HKP.05.00/2020, and partial support from MIT-Indonesia Research Alliance (MIRA) managed by Institut Teknologi Bandung (ITB) through WCU Program.

Conflicts of Interest: The authors declare no conflict of interest.

References

1. World Health Organization (WHO). (27 April 2020). WHO Timeline-COVID-19. Available online: https://www.who.int/news-room/detail/27-04-2020-who-timeline---covid-19 (accessed on 27 May 2020).

2. Kampf, G.; Todt, D.; Pfaender, S.; Steinmann, E. Persistence of Coronaviruses on Inanimate Surfaces and Their Inactivation with Biocidal Agents. *J. Hosp. Infect.* **2020**, *104*, 246–251. [CrossRef] [PubMed]

3. Rai, H.; Knighton, S.; Zabarsky, T.F.; Donskey, C.J. Comparison of Ethanol Hand Sanitizer Versus Moist Towelette Packets for Mealtime Patient Hand Hygiene. *Am. J. Infect. Control.* **2017**, *45*, 1033–1034. [CrossRef] [PubMed]

4. Foddai, A.C.G.; Grant, I.R.; Dean, M. Efficacy of Instant Hand Sanitizers against Foodborne Pathogens Compared with Hand Washing with Soap and Water in Food Preparation Settings: A Systematic Review. *J. Food Prot.* **2016**, *79*, 1040–1054. [CrossRef] [PubMed]

5. Jing, J.L.J.; Yi, T.P.; Bose, R.J.C.; McCarthy, J.R.; Tharmalingam, N.; Madheswaran, T. Hand Sanitizers: A Review on Formulation Aspects, Adverse Effects, and Regulations. *Int. J. Environ. Res. Public Health* **2020**, *17*, 3326. [CrossRef] [PubMed]

6. Golin, A.P.; Choi, D.; Ghahary, A. Hand Santizers: A Review of Ingredients, Mechanisms of Action, Modes of Delivery, and Efficacy against Coronaviruses. *Am. J. Infect. Control.* **2020**, in press. [CrossRef]

7. WHO. (April 2010). Guide to Local Production: WHO-Recommended Handrub Formulations. Available online: https://www.who.int/gpsc/5may/Guide_to_Local_Production.pdf?ua=1 (accessed on 26 May 2020).

8. Rastogi, M.; Shrivastava, S. Recent Advances in Second Generation Bioethanol Production: An Insight to Pretreatment, Saccharification and Fermentation Processes. *Renew. Sustain. Energy Rev.* **2017**, *80*, 330–340. [CrossRef]

9. Rosales-Calderon, O.; Arantes, V. A Review on Commercial-Scale High-Value Products That Can be Produced alongside Cellulosic Ethanol. *Biotechnol. Biofuels* **2019**, *12*, 240. [CrossRef]

10. Su, T.; Zhao, D.; Khodadadi, M.; Len, C. Lignocellulosic Biomass for Bioethanol: Recent Advances, Technology Trends, and Barriers to Industrial Development. *Curr. Opin. Green Sustain. Chem.* **2020**, *24*, 56–60. [CrossRef]

11. Samsuri, M.; Gozan, M.; Mardias, R.; Baiquni, M.; Hermansyah, H.; Wijanarko, A.; Prasetya, B.; Nasikin, M. Pemanfaatan Sellulosa Bagas untuk Produksi Ethanol Melalui Sakarifikasi dan Fermentasi Serentak dengan Enzim Xylanase. *Makara* **2007**, *11*, 17–24. [CrossRef]

12. Arlofa, N.; Gozan, M.; Pradita, T.; Jufri, M. Optimization of Bioethanol Production from Durian Skin by Encapsulation of Saccharomyces cerevisiae. *Asian J. Chem.* **2019**, *31*, 1027–1033. [CrossRef]

13. Darmawan, M.A.; Hermawan, Y.A.; Samsuri, M.; Gozan, M. Conversion of Paper Waste to Bioethanol using Selected Enzyme Combination (Cellulase and Cellobiase) Through Simultaneous Saccharification and Fermentation. In Proceedings of the 11th Regional Conference on Chemical Engineering (RCChE 2018), Yogyakarta, Indonesia, 7–8 November 2019; Teguh, A., Imam, P., Rochmadi, N., Rofiqoh Eviana, P., Eds.; American Institute of Physics Inc.: College Park, MD, USA, 2019.

14. Sahlan, M.; Hermansyah, H.; Wijanarko, A.; Gozan, M.; Lischer, K.; Ahmudi, A.; Pujianto, P. Ethanol Production by Encapsulated Rhyzopus oryzae from Oil Palm Empty Fruit Bunch. *Evergreen* **2020**, *7*, 92–96. [CrossRef]

15. Koppram, R.; Nielsen, F.; Albers, E. Simultaneous Saccharification and Co-Fermentation for Bioethanol Production using Corncobs at Lab, PDU and Demo Scales. *Biotechnol. Biofuels* **2013**, *6*. [CrossRef] [PubMed]

16. Khuong, L.D.; Kondo, R.; De Leon, R.; Anh, T.K.; Shimizu, K.; Kamei, I. Bioethanol Production from Alkaline-Pretreated Sugarcane Bagasse by Consolidated Bioprocessing using *Phlebia* sp. MG-60. *Int. Biodeterior. Biodegrad.* **2014**, *88*, 62–68. [CrossRef]

17. Lennartsson, P.R.; Erlandsson, P.; Taherzadeh, M.J. Integration of the First and Second Generation Bioethanol Processes and the Importance of By-Products. *Bioresour. Technol.* **2014**, *165*, 3–8. [CrossRef] [PubMed]

18. Sharma, B.; Larroche, C.; Dussap, C.G. Comprehensive Assessment of 2G Bioethanol Production. *Bioresour. Technol.* **2020**, *313*, 1–9. [CrossRef] [PubMed]

19. Shinoj, S.; Visvanathan, R.; Panigrahi, S.; Kochubabu, M. Oil Palm Fiber (OPF) and Its Composites: A Review. *Ind. Crops Prod.* **2011**, *33*, 7–22. [CrossRef]

20. Purwandari, F.A.; Sanjaya, A.P.; Millati, R.; Cahyanto, M.N.; Horváth, I.S.; Niklasson, C.; Taherzadeh, M.J. Pretreatment of Oil Palm Empty Fruit Bunch (OPEFB) by N-methylmorpholine-N-oxide (NMMO) for Biogas Production: Structural Changes and Digestion Improvement. *Bioresour. Technol.* **2013**, *128*, 461–466. [CrossRef]

21. Medina, J.D.C.; Woiciechowski, A.; Filho, A.Z.; Nigam, P.S.; Ramos, L.P.; Soccol, C.R. Steam Explosion Pretreatment of Oil Palm Empty Fruit Bunches (EFB) using Autocatalytic Hydrolysis: A Biorefinery Approach. *Bioresour. Technol.* in press. [CrossRef]

22. Panjaitan, J.R.H.; Gozan, M. Formic Acid Production from Palm Oil Empty Fruit Bunches. *Int. J. Appl. Eng. Res.* **2017**, *12*, 4382–4390.

23. Harahap, A.F.P.; Rahman, A.A.; Sadrina, I.N.; Gozan, M. Production of Formic Acid from Oil Palm Empty Fruit Bunch via Dilute Acid Hydrolysis by Response Surface Methodology. In Proceedings of the Broad Exposure to Science and Technology 2019 (BEST2019), Bali, Indonesia, 7–8 August 2019; IOP Publishing Ltd.: Bristol, UK, 2019.

24. Gozan, M.; Panjaitan, J.R.H.; Tristantini, D.; Alamsyah, R.; Yoo, Y.J. Evaluation of Separate and Simultaneous Kinetic Parameters for Levulinic Acid and Furfural Production from Pretreated Palm Oil Empty Fruit Bunches. *Int. J. Chem. Eng.* **2018**, *2018*, 1–12. [CrossRef]

25. McInnes, A. *A Comparison of Leading Palm Oil Certification Standards*; Forest Peoples Programme: Moreton-in-Marsh, UK, 2017; pp. 5–37.

26. Bernhard, T. Definition and Scope of Systems Engineering. In *Systems Engineering: Principles and Practice of Computer-Based Systems Engineering*; John Wiley & Sons Ltd.: Chichester, West Sussex, UK, 1993; pp. 1–23.

27. Dimian, A.; Bildea, C.; Kiss, A. *Integrated Design and Simulation of Chemical Processes*, 2nd ed.; Elsevier Science: Amsterdam, The Netherland, 2014; pp. 35–71.

28. Feibel, B.J. *Investment Performance Measurement*, 1st ed.; John Wiley & Sons, Inc.: Hoboken, NJ, USA, 2003; pp. 215–298.

29. Khor, C.; Elkamel, A.; Anderson, W. Incorporating the Systems Approach in Future Undergraduate Chemical Engineering Education Curriculum: Illustration via Computer-Aided Process Simulation Laboratory Exercises. *World Rev. Sci. Technol. Sustain. Dev.* **2008**, *5*, 402–413. [CrossRef]

30. Seider, W.D.; Seader, J.D.; Lewin, D.R.; Widagdo, S. *Product and Process Design Principles: Synthesis, Analysis, and Evaluation*, 3rd ed.; John Wiley & Sons, Inc.: New York, NY, USA, 2004; pp. 110–147.

31. Zhang, K.; Pei, Z.; Wang, D. Organic Solvent Pretreatment of Lignocellulosic Biomass for Biofuels and Biochemicals: A Review. *Bioresour. Technol.* **2016**, *199*, 21–33. [CrossRef] [PubMed]

32. Zulkiple, N.; Maskat, M.Y.; Hassan, O. Pretreatment of Oil Palm Empty Fruit Fiber (OPEFB) with Aqueous Ammonia for High Production of Sugar. *Procedia Chem.* **2016**, *18*, 155–161. [CrossRef]

33. Badan Standardisasi Nasional (BSN). *Standar Nasional Indonesia (Indonesian National Standardization) SNI 0444: 2009 Pulp—Cara Uji Kadar Selulosa Alfa, Beta dan Gamma*; BSN: Jakarta, Indonesia, 2009.

34. Badan Standardisasi Nasional (BSN). *Standar Nasional Indonesia (Indonesian National Standardization) SNI 14-1304-1989 Cara Uji Kadar Pentosan Pulp Kayu*; BSN: Jakarta, Indonesia, 1989.

35. Badan Standardisasi Nasional (BSN). *Standar Nasional Indonesia (Indonesian National Standardization) SNI 0492:2008 Pulp dan Kayu—Cara Uji Kadar Lignin-Metode Klason*; BSN: Jakarta, Indonesia, 2008.

36. Badan Standardisasi Nasional (BSN). *Standar Nasional Indonesia (Indonesian National Standardization) SNI 0442: 2009 Kertas, Karton, dan Pulp-Cara Uji Kadar Abu pada 525 °C*; BSN: Jakarta, Indonesia, 2009.

37. Fogler, H.S. *Elements of Chemical Reaction Engineering*, 4th ed.; Prentice Hall: Laflin, PA, USA, 2006; pp. 91–98.

38. Fengel, D.; Wegener, G. *Wood: Chemistry, Ultrastructure, Reactions*, 2nd ed.; Walter de Gruyter: Berlin, Germany, 1989; pp. 162–197.

39. Amenaghawon, N.A.; Okieimen, C.O.; Ogbeide, S.E. Kinetic Modelling of Ethanol Inhibition during Alcohol Fermentation of Corn Stover using Saccharomyces cerevisiae. *Int. J. Eng. Res. Appl.* **2012**, *2*, 798–803.

40. Dodic, J.M.; Vucurovic, D.G.; Dodic, S.N.; Grahovac, J.A.; Popov, S.D.; Nedeljkovic, N.M. Kinetic Modelling of Batch Ethanol Production from Sugar Beet Raw Juice. *Appl. Energy* **2012**, *99*, 192–197. [CrossRef]

41. Phisalaphong, M.; Srirattana, N.; Tanthapanichakoon, W. Mathematical Modeling to Investigate Temperature Effect on Kinetic Parameters of Ethanol Fermentation. *Biochem. Eng. J.* **2006**, *28*, 36–43. [CrossRef]

42. Peters, M.S.; Timmerhaus, K.D. *Plant Design and Economics for Chemical Engineers*, 4th ed.; McGraw-Hill: New York, NY, USA, 1991; pp. 120–252.

43. Konda, N.M.; Shi, J.; Singh, S.; Blanch, H.W.; Simmons, B.A.; Marcuschamer, D.K. Understanding Cost Drivers and Economic Potential of Two Variants of Ionic Liquid Pretreatment for Cellulosic Biofuel Production. *Biotechnol. Biofuels* **2014**, *7*, 1–11. [CrossRef]

44. Damodaran, A. (January 2020). Margins by Sector (US). Available online: http://pages.stern.nyu.edu/~adamodar/New_Home_Page/datafile/margin.html (accessed on 10 July 2020).

MDPI

St. Alban-Anlage 66

4052 Basel

Switzerland

Tel. +41 61 683 77 34

Fax +41 61 302 89 18

www.mdpi.com

Applied Sciences Editorial Office

E-mail: applsci@mdpi.com

www.mdpi.com/journal/applsci